Graduate Texts in Mathematics 162

Springer
New York
Berlin
Heidelberg
Barcelona
Budapest
Hong Kong
London
Milan
Paris
Tokyo

Graduate Texts in Mathematics

continued after index

J.L. Alperin
with Rowen B. Bell

Groups and
Representations

 Springer

J.L. Alperin
Rowen B. Bell
Department of Mathematics
University of Chicago
Chicago, IL 60637-1514

Mathematics Subject Classifications (1991): 20-01

Library of Congress Cataloging-in-Publication Data
Alperin, J.L.
    Groups and representations / J.L. Alperin with Rowen B. Bell.
        p. cm. — (Graduate texts in mathematics ; 162)
    Includes bibliographical references (p.   —) and index.
    ISBN 0-387-94525-3 (alk. paper). — ISBN 0-387-94526-1
    (pbk.: alk. paper)
        1. Representations of groups. I. Bell, Rowen B.   II. Title.
    III. Series.
    QA176.A46   1995
    512′.2 — dc20   95-17160

Printed on acid-free paper.

Production managed by Robert Wexler; manufacturing supervised by Jeffrey Taub.
Photocomposed copy prepared from the author's LaTeX file.
Printed and bound by R.R. Donnelley & Sons, Harrisonburg, VA.
Printed in the United States of America.

9 8 7 6 5 4 3 2 1

ISBN 0-387-94525-3 Springer-Verlag New York Berlin Heidelberg (Hardcover)
ISBN 0-387-94526-1 Springer-Verlag New York Berlin Heidelberg (Softcover)

Preface

This book is based on a first-year graduate course given regularly by the first author at the University of Chicago, most recently in the autumn quarters of 1991, 1992, and 1993. The lectures given in this course were expanded and prepared for publication by the second author.

The aim of this book is to provide a concise yet thorough treatment of some topics from group theory and representation theory with which every mathematician should be well acquainted. Of course, the topics covered naturally reflect the viewpoints and interests of the authors; for instance, we make no mention of free groups, and the emphasis throughout is admittedly on finite groups. Our hope is that this book will enable graduate students from every mathematical field, as well as bright undergraduates with an interest in algebra, to solidify their knowledge of group theory.

As the course on which this book is based is required for all incoming mathematics graduate students at Chicago, we make very modest assumptions about the algebraic background of the reader. A nodding familiarity with groups, rings, and fields, along with some exposure to elementary number theory and a solid knowledge of linear algebra (including, at times, familiarity with canonical forms of matrices), should be sufficient preparation.

We now give a brief summary of the book's contents. The first four chapters are devoted to group theory. Chapter 1 contains a review (largely without proofs) of the basics of group theory, along with material on automorphism groups, semidirect products, and group actions. These latter concepts are among our primary tools in the book and are often not covered adequately during one's first exposure to group theory. Chapter 2 discusses the structure of the general linear groups and culminates with a proof of the simplicity of the projective special linear groups. An understanding of this material is an essential (but often overlooked) component of any substantive study of group theory; for, as the first author once wrote:

> The typical example of a finite group is $GL(n, q)$, the general linear group of n dimensions over the field with q elements. The student who is introduced to the subject with other examples is being completely misled. [3, p. 121]

Chapter 3 concentrates on the examination of finite groups through their p-subgroups, beginning with Sylow's theorem and moving on to such results as the Schur-Zassenhaus theorem. Chapter 4 starts with the Jordan-Hölder theorem and continues with a discussion of solvable and nilpotent groups. The final two chapters focus on finite-dimensional algebras and the representation theory of finite groups. Chapter 5 is centered around Maschke's theorem and Wedderburn's structure theorems for semisimple algebras. Chapter 6 develops the ordinary character theory of finite groups, including induced characters, while the Appendix treats some additional topics in character theory that require a somewhat greater algebraic background than does the core of the book.

We have included close to 200 exercises, and they form an integral part of the book. We have divided these problems into "exercises" and "further exercises;" the latter category is generally reserved for exercises that introduce and develop theoretical concepts not included in the text. The level of the problems varies from routine to difficult, and there are a few that we do not expect any student to be able to handle. We give no indication of the degree of difficulty of each exercise, for in mathematical research one does not know in advance what amount of work will be required to complete any step! In an effort to keep our exposition self-contained, we have strived to keep references in the text to the exercises at a minimum.

The sections of this book are numbered continuously, so that Section 4 is actually the first section of Chapter 2, and so forth. A citation of the form "Proposition Y" refers to the result of that name in the current section, while a citation of the form "Proposition X.Y" refers to Proposition Y of Section X.

We would like to extend our thanks to: Michael Maltenfort and Colin Rust, for their thought-provoking proofreading and their many constructive suggestions during the preparation of this book; the students in the first author's 1993 course, for their input on an earlier draft of this book which was used as that course's text; Efim Zelmanov and the students in his 1994 Chicago course, for the same reason; and the University of Chicago mathematics department, for continuing to provide summer support for graduate students, as without such support this book would not have been written in its present form. We invite you to send notice of errors, typographical or otherwise, to the second author at `bell@math.uchicago.edu`.

In remembrance of a life characterized by integrity, devotion to family, and service to community, the second author would like to dedicate this book to David Wellman (1953–1995).

Contents

1
Rudiments of Group Theory

In this introductory chapter, we review the elementary notions of group theory and develop many of the tools that we will use in the remaining chapters. Section 1 consists primarily of those facts with which we assume the reader is familiar from some prior study of group theory; consequently, most proofs in this section have been omitted. In Section 2 we introduce some important concepts, such as automorphism groups and semidirect products, which are not necessarily covered in a first course on group theory. Section 3 treats the theory of group actions; here we present both elementary applications and results of a more technical nature which will be needed in later chapters.

1. Review

Recall that a *group* consists of a non-empty set G and a binary operation on G, usually written as multiplication, satisfying the following conditions:

- The binary operation is associative: $(xy)z = x(yz)$ for any $x, y, z \in G$.
- There is a unique element $1 \in G$, called the *identity element* of G, such that $x1 = x$ and $1x = x$ for any $x \in G$.

- For every $x \in G$ there is a unique element $x^{-1} \in G$, called the *inverse* of x, with the property that $xx^{-1} = 1$ and $x^{-1}x = 1$.

Associativity allows us to consider unambiguously the product of any finite number of elements of a group. The order of the elements in such a product is critically important, for if x and y are elements of a group G, then it is not necessarily true that $xy = yx$. If this happens, then we say that x and y *commute*. More generally, we define the *commutator* of x and y to be the element $[x, y] = xyx^{-1}y^{-1}$, so that x and y commute iff $[x, y] = 1$. (Many authors define $[x, y] = x^{-1}y^{-1}xy$.) We say that G is *abelian* if all pairs of elements of G commute, in which case the order of elements in a product is irrelevant; otherwise, we say that G is *non-abelian*. The group operation of an abelian group may be written additively, meaning that the product of elements x and y is written as $x + y$ instead of xy, the inverse of x is denoted by $-x$, and the identity element is denoted by 0.

If x is an element of a group G, then for $n \in \mathbb{N}$ we use x^n (resp., x^{-n}) to mean the product $x \cdots x$ (resp., $x^{-1} \cdots x^{-1}$) involving n terms. We also define $x^0 = 1$. (In an abelian group that is written additively, we write nx instead of x^n for $n \in \mathbb{Z}$.) It is easily seen that the usual rules for exponentiation hold. We say that x is of *finite order* if there is some $n \in \mathbb{N}$ such that $x^n = 1$. If x is of finite order, then we define the *order* of x to be the least positive integer n such that $x^n = 1$. Clearly, x is of order n iff $1, x, x^2, \ldots, x^{n-1}$ are distinct elements of G and $x^n = 1$.

A group G is said to be *finite* if it has a finite number of elements, and *infinite* otherwise. We define the *order* of a finite group G, denoted $|G|$, to be the number of elements of G; we may also use $|S|$ for the cardinality of any finite set S. Every element of a finite group is of finite order, and there are infinite groups with this property; these groups are said to be *periodic*. However, there are infinite groups in which the identity element is the only element of finite order; such groups are said to be *torsion-free*.

A subset H of a group G is said to be a *subgroup* of G if it forms a group under the restriction to H of the binary operation on G. Equivalently, $H \subseteq G$ is a subgroup iff the following conditions hold:

- The identity element 1 of G lies in H.
- If $x, y \in H$, then their product xy in G lies in H.
- If $x \in H$, then its inverse x^{-1} in G lies in H.

Clearly G is a subgroup of itself. The set $\{1\}$ is also a subgroup of G; it is called the *trivial subgroup*, and for the sake of simplicity we denote it by 1. Every subgroup of a finite group is finite; however, an infinite group always has both finite and infinite subgroups, namely its trivial subgroup and itself, respectively. Similarly, every subgroup of an abelian group is abelian, but a non-abelian group always has both abelian and non-abelian subgroups. If H is a subgroup of G, then we write $H \leqslant G$; if H is properly contained in G, then we call H a *proper subgroup* of G, and we may write $H < G$. (This notational distinction is common, but not universal.) If $K \leqslant H$ and $H \leqslant G$, then evidently $K \leqslant G$.

PROPOSITION 1. If H and K are subgroups of a group G, then so is their intersection $H \cap K$. More generally, the intersection of any collection of subgroups of a group is also a subgroup of that group. ■

The following theorem gives important information about the nature of subgroups of a finite group.

LAGRANGE'S THEOREM. Let G be a finite group, and let $H \leqslant G$. Then $|H|$ divides $|G|$. ■

If X is a subset of a group G, then we define $<X>$ to be the intersection of all subgroups of G which contain X. By Proposition 1, $<X>$ is a subgroup of G, which we call the *subgroup of G generated by X*. We see that $<X>$ is the smallest subgroup of G which contains X, in the sense that it is contained in any such subgroup; hence if $X \leqslant G$, then $<X> = X$. If $X = \{x\}$, then we write $<x>$ in lieu of $<X>$; similarly, if $X = \{x_1, \ldots, x_n\}$, then we write $<x_1, \ldots, x_n>$ for $<X>$.

PROPOSITION 2. Let X be a subset of a group G. Then $<X>$ consists of the identity and all products of the form $x_1^{\epsilon_1} \cdots x_r^{\epsilon_r}$ where $r \in \mathbb{N}$, $x_i \in X$, and $\epsilon_i = \pm 1$ for all i. ■

A group G is said to be *cyclic* if $G = <g>$ for some $g \in G$; the element g is called a *generator* of G. For example, if G is a group of order n having an element g of order n, then $G = <g>$ since $g, \ldots, g^{n-1}, g^n = 1$ are n distinct elements of G. By Proposition 2 we have $<g> = \{g^n \mid n \in \mathbb{Z}\}$, and consequently we see via the exponentiation relations that cyclic groups are abelian; nonetheless, we will generally write cyclic groups multiplicatively instead of additively. If g is of order n, then $<g> = \{1, g, \ldots, g^{n-1}\}$, and hence $|<g>| = n$. If g is not of finite order, then $<g>$ is a torsion-free infinite abelian group. Any two finite cyclic groups of the same order are "equivalent" in a sense that will be made precise later in this section, and any two infinite cyclic groups are equivalent in the same sense. The canonical infinite cyclic group is $\mathbb{Z}$, the set of integers under addition, while the canonical cyclic group of order n is $\mathbb{Z}/n\mathbb{Z}$, the set of residue classes of the integers under addition modulo n.

Suppose that G is a finite group and $g \in G$ is of order n. Then $<g>$ is a subgroup of G of order n, so by Lagrange's theorem we see that n divides $|G|$. Thus, the order of an element of a finite group must divide the order of that group. Consequently, if $|G|$ is equal to some prime p, then the order of each element of G must be a non-trivial divisor of p, from which it follows that G is cyclic with every non-identity element of G being a generator.

If X and Y are subsets of a group G, then we define the *product* of X and Y in G to be $XY = \{xy \mid x \in X, y \in Y\} \subseteq G$. We can extend this definition to any finite number of subsets of G. We also define the *inverse* of $X \subseteq G$ by $X^{-1} = \{x^{-1} \mid x \in X\} \subseteq G$. If H is a non-empty subset of G, then $H \leqslant G$ iff $HH = H$ and $H^{-1} = H$.

PROPOSITION 3. Let H and K be subgroups of a group G. Then HK is a subgroup of G iff $HK = KH$. ∎

Observe that if H and K are subgroups of G, then their product HK contains both H and K; if in addition $K \leqslant H$, then $HK = H$. (These properties do not hold if H and K are arbitrary subsets of G.) If G is abelian, then $HK = KH$ for any subgroups H and K of G, and hence the product of any two subgroups of an abelian group is a subgroup.

We can now describe the subgroup structure of finite cyclic groups.

THEOREM 4. Let $G = <g>$ be a cyclic group of order n. Then:

(i) For each divisor d of n, there is exactly one subgroup of G of order d, namely $<g^{\frac{n}{d}}>$.

(ii) If d and e are divisors of n, then the intersection of the subgroups of orders d and e is the subgroup of order $\gcd(d, e)$.

(iii) If d and e are divisors of n, then the product of the subgroups of orders d and e is the subgroup of order $\text{lcm}(d, e)$. ∎

If $H \leqslant G$ and $x \in G$, then we write xH instead of $\{x\}H$; the set xH is called a *left coset* of H in G. Similarly, we write Hx instead of $H\{x\}$, and we call Hx a *right coset* of H in G. In this book we shall use left cosets, and consequently from now on the word "coset" should be read as "left coset." Our use of left cosets instead of right cosets is essentially arbitrary, as any statement that we make about left cosets has a valid counterpart involving right cosets. Indeed, many group theory texts use right cosets where we use left cosets. There is a bijective correspondence between left and right cosets of H in G, sending a left coset xH to its inverse $(xH)^{-1} = Hx^{-1}$.

Let H be a subgroup of G. Any two cosets of H in G are either equal or disjoint, with cosets xH and yH being equal iff $y^{-1}x \in H$. Consequently, an element $x \in G$ lies in exactly one coset of H, namely xH. For any $x \in G$, there is a bijective correspondence between H and xH; one such correspondence sends $h \in H$ to xh. We define the *index* of H in G, denoted $|G : H|$, to be the number of cosets of H in G. (If there is an infinite number of cosets of H in G, then we could define $|G : H|$ to be the appropriate cardinal number without changing the truth of any statements made below, as long as we redefine $|G|$ as being the cardinal number $|G : 1|$.) The cosets of H in G partition G into $|G : H|$ disjoint sets of cardinality $|H|$, and hence we have $|G| = |G : H||H|$. (This observation proves Lagrange's theorem; however, it is possible to prove Lagrange's theorem without reference to cosets by means of a simple counting argument.) In particular, all subgroups of a finite group are of finite index, while subgroups of an infinite group may be of finite or infinite index. We denote the set of cosets (or the *coset space*) of H in G by G/H.

We can now give a complete description of the subgroups of infinite cyclic groups. We invite the reader to restate Theorem 4 in such a way so as to make the parallelism between Theorems 4 and 5 more explicit.

THEOREM 5. Let $G = <g>$ be an infinite cyclic group. Then:

  (i) For each $d \in \mathbb{N}$, there is exactly one subgroup of G of index d, namely $<g^d>$. Furthermore, every non-trivial subgroup of G is of finite index.

 (ii) Let $d, e \in \mathbb{N}$. Then the intersection of the subgroups of indices d and e is the subgroup of index $\operatorname{lcm}(d, e)$.

(iii) Let $d, e \in \mathbb{N}$. Then the product of the subgroups of indices d and e is the subgroup of index $\gcd(d, e)$. ∎

The following result generalizes Lagrange's theorem and shall be referred to as "factorization of indices."

THEOREM 6. *If $K \leqslant H \leqslant G$, then $|G : K| = |G : H||H : K|$.* ∎

Let H be a subgroup of a group G, and let $\mathcal{I}$ be an indexing set that is in bijective correspondence with the coset space of H in G. A subset $T = \{t_i \mid i \in \mathcal{I}\}$ of G is said to be a *(left) transversal* for H (or a set of *(left) coset representatives* of H in G) if the sets $t_i H$ are precisely the cosets of H in G, with no coset omitted or duplicated.

Let N be a subgroup of a group G. We say that N is a *normal subgroup* of G (or that N is *normal* in G) if $xN = Nx$ for all $x \in G$, or equivalently if $xNx^{-1} \subseteq N$ for all $x \in G$. If G is abelian, then every subgroup of G is normal. The subgroups 1 and G are always normal in G; if these are the only normal subgroups of G, then we say that G is *simple*. For example, a cyclic group of prime order is simple. (A group having only one element is by convention not considered to be simple.) If N is normal in G, then we write $N \trianglelefteq G$; if N is both proper and normal in G, then we may write $N \triangleleft G$. (Once again, many authors do not make this distinction and instead use $N \triangleleft G$ to mean simply that N is normal in G.) If $H \trianglelefteq G$ and $K \trianglelefteq H$, then it is not necessarily true that $K \trianglelefteq G$; we will provide a counterexample momentarily. However, it is clearly true that if $K \trianglelefteq G$ and $K \leqslant H \leqslant G$, then $K \trianglelefteq H$.

PROPOSITION 7. *Let H and K be subgroups of a group G. If $K \trianglelefteq G$, then $HK \leqslant G$ and $H \cap K \trianglelefteq H$; if also $H \trianglelefteq G$, then $HK \trianglelefteq G$ and $H \cap K \trianglelefteq G$.* ∎

PROPOSITION 8. *Any subgroup of index 2 is normal.*

PROOF. Let $H \leqslant G$, and suppose that $|G : H| = 2$. Then there are two left cosets of H in G; one is H, and thus the other must be $G - H$. Similarly, H and $G - H$ are the two right cosets of H in G. It now follows that $x \in H$ iff $xH = H = Hx$, and $x \notin H$ iff $xH = G - H = Hx$; hence $H \trianglelefteq G$. ∎

Normal subgroups are important because they allow us to create new groups from old, in the following way:

THEOREM 9. *If $N \trianglelefteq G$, then the coset space G/N forms a group under the binary operation defined by $(xN)(yN) = (xy)N$.* ∎

If $N \trianglelefteq G$, then we call G/N with the above binary operation the *quotient group* of G by N. The identity element of G/N is N, and the inverse of $xN \in G/N$ is $x^{-1}N$. If G is abelian, then G/N is also abelian.

Let x and g be elements of a group G. The *conjugate* of x by g is defined to be the element gxg^{-1} of G. (Some authors define the conjugate of x by g to be $g^{-1}xg$. The notations ^{g}x and x^{g} are sometimes used for gxg^{-1} and $g^{-1}xg$, respectively.) Two elements x and y of G are said to be *conjugate* if there exists some $g \in G$ such that $y = gxg^{-1}$. No two distinct elements of an abelian group can be conjugate. A subgroup N of G is normal iff every conjugate of an element of N by an element of G lies in N.

Let X be a set. A *permutation* of X is a bijective set map from X to X. The set of permutations of X, denoted Σ_X, forms a group under composition of mappings. If $X = \{1, \ldots, n\}$ for some $n \in \mathbb{N}$, then this group is called the *symmetric group of degree n* and is denoted Σ_n. (Many authors denote this group by S_n or $\mathfrak{S}_n$.) The group Σ_n is finite and of order $n! = n(n-1) \cdots 2 \cdot 1$.

An element ρ of Σ_n is called a *cycle of length r* (or an *r-cycle*) if there are distinct integers $1 \le a_1, \ldots, a_r \le n$ such that $\rho(a_i) = (a_{i+1})$ for all $1 \le i < r$, $\rho(a_r) = a_1$, and $\rho(b) = b$ for any $1 \le b \le n$ which is not equal to some a_i. If the cycle ρ is as defined above, then we write $\rho = (a_1 \cdots a_r)$. Of course, this can be done in r different ways; for example, (1 2 4), (2 4 1), and (4 1 2) denote the same 3-cycle in Σ_4. The cycle ρ as defined above is said to *move* each a_i and *fix* every other number. Two cycles are said to be *disjoint* if there is no number that is moved by both cycles. The product of two cycles $(a_1 \cdots a_r)$ and $(b_1 \cdots b_s)$ is written $(a_1 \cdots a_r)(b_1 \cdots b_s)$; if $a_i = b_j$, then this product moves b_{j-1} to a_{i+1}. (We read from "right to left" in this manner because we think of the cycles as being functions on $\{1, \ldots, n\}$, and so the product of two cycles corresponds to a composition of functions, which we choose to perform from right to left in the usual fashion. In many group theory texts, composition is performed from left to right.)

Every element of Σ_n can be written as a product of disjoint cycles; such an expression is called a *disjoint cycle decomposition* of the permutation. Any two disjoint cycle decompositions of a given permutation must necessarily include the same cycles, but possibly in some different order. Therefore we can associate, in a well-defined way, a collection of positive integers whose sum is n to each element ρ of Σ_n; this partition of n consists of the lengths of the cycles that appear in a disjoint cycle decomposition of ρ and is called the *cycle structure* of ρ. For example, the cycle structure of an r-cycle in Σ_n is the partition $(r, 1, \ldots, 1)$ having $n - r$ ones; the cycle structure of $(1\ 2\ 4)(3\ 5)$ in Σ_6 is the partition $(3, 2, 1)$. We generally omit 1-cycles when writing a permutation as a product of disjoint cycles. As usual, we will use 1 to denote the identity element of Σ_n, whose disjoint cycle decomposition consists solely of 1-cycles.

PROPOSITION 10. Let $n \in \mathbb{N}$. Then two elements of Σ_n are conjugate iff they have the same cycle structure. ∎

For a proof, see [24, pp. 46–7].

A *transposition* in Σ_n is a 2-cycle. Every element of Σ_n can be written as a (not necessarily disjoint) product of transpositions in many different ways. However, it can be shown that any two expressions of a given permutation as a product of transpositions use the same number, modulo 2, of transpositions. (See [24, pp. 8–9].) Hence we can say that a permutation is *even* (resp., *odd*) if it can be written as a product of an even (resp., odd) number of transpositions, for a permutation is either even or odd, but never both. For example, since an r-cycle can be written as a product of $r - 1$ transpositions, we see that a cycle is an even permutation iff its length is odd. The subset of Σ_n consisting of all even permutations is a subgroup of index 2, and hence is normal in Σ_n by Proposition 8; it is called the *alternating group of degree n* and is denoted A_n.

Consider $H = \{1, (1\ 2)(3\ 4), (1\ 3)(2\ 4), (1\ 4)(2\ 3)\} \subseteq A_4$. One can show that $H \trianglelefteq A_4$. (In fact, H is normal in Σ_4. This group H is historically called the *Klein four-group*.) Let $K = \{1, (1\ 2)(3\ 4)\}$. Then K is a subgroup of H with $|H : K| = |H|/|K| = 4/2 = 2$, and hence $K \trianglelefteq H$ by Proposition 8. However, by conjugating $(1\ 2)(3\ 4)$ by the even permutation $(1\ 2\ 3)$, we see that K is not normal in A_4. This provides the counterexample referred to on page 6.

Let G and H be groups. A *homomorphism* is a map $\varphi: G \to H$ with the property that $\varphi(xy) = \varphi(x)\varphi(y)$ for all $x, y \in G$; that is, a homomorphism is a map between groups which preserves the respective group structures. If φ is a homomorphism, then $\varphi(1) = 1$, and $\varphi(x^{-1}) = \varphi(x)^{-1}$ for any element x. The *trivial homomorphism* from G to H is the map sending every element of G to the identity element of H. If a homomorphism φ is injective, then we call φ a *monomorphism*, and if φ is surjective, we call φ an *epimorphism*; we say that φ is an *isomorphism* if φ is bijective. (Recall that a set map $f: X \to Y$ is called injective if $f(x) = f(x')$ forces $x = x'$, surjective if for any $y \in Y$ we have $f(x) = y$ for some $x \in X$, and bijective if it is both injective and surjective.) If φ is an isomorphism, then so is $\varphi^{-1}: H \to G$. A homomorphism $\varphi: G \to G$ is called an *endomorphism* of G; a bijective endomorphism is called an *automorphism*.

If G and H are groups and there is an isomorphism $\varphi: G \to H$, then we say that G and H are *isomorphic*, or that G is isomorphic with H, and we write $G \cong H$. The notion of isomorphism is an equivalence relation on groups; that is, it is reflexive ($G \cong G$), symmetric ($G \cong H$ implies $H \cong G$), and transitive ($G \cong H$ and $H \cong K$ together imply $G \cong K$). Therefore, we can speak of the "isomorphism class" to which a given group belongs. Isomorphic groups are to be thought of as being virtually identical, in the sense that any statement made about a group is true (after making appropriate identifications) for any other group with which it is isomorphic. If we say that a group having certain properties is "unique," then we often mean that it is "unique up to isomorphism," by which we mean that any two groups having the specified properties are isomorphic.

We now consider some standard examples.

- Let $G = <g>$ and $H = <h>$ be two cyclic groups of order n. We define a map $\varphi: G \to H$ by setting $\varphi(g^a) = h^a$ for every $0 \leq a < n$. This map φ is an isomorphism. Consequently, any two finite cyclic groups of the same order are isomorphic. In particular, any cyclic group of order n is isomorphic with $\mathbb{Z}/n\mathbb{Z}$, and there is a unique group of order p for each prime p. We will use $\mathbf{Z_n}$ to denote a cyclic group of order n, written multiplicatively. We can similarly show that any two infinite cyclic groups are isomorphic; we will use $\mathbf{Z}$ to denote an infinite cyclic group, written multiplicatively.

- Let G be a group, let $H \leqslant G$, and let $g \in G$. The *conjugate* of H by g is the set $gHg^{-1} = \{ghg^{-1} \mid h \in H\}$ consisting of all conjugates of elements of H by g. It is easily verified that $gHg^{-1} \leqslant G$. We say that $K \leqslant G$ is a *conjugate* of H

in G, or that K and H are conjugate in G, if $K = gHg^{-1}$ for some $g \in G$. Given $H \leqslant G$ and $g \in G$, we define a map $\varphi \colon H \to gHg^{-1}$ by $\varphi(h) = ghg^{-1}$ for $h \in H$. We see easily that φ is an isomorphism; hence conjugate subgroups are isomorphic. However, it is not true that any two isomorphic subgroups of a group G are conjugate in G. For example, the Klein four-group has three subgroups of order 2 which are necessarily isomorphic but which, being subgroups of an abelian group, cannot be conjugate.

- Let $X = \{x_1, \ldots, x_n\}$ and let Σ_X be the group of permutations of X. We define a map $\varphi \colon \Sigma_n \to \Sigma_X$ by $\varphi(\rho)(x_i) = x_{\rho(i)}$ for $\rho \in \Sigma_n$ and $1 \leq i \leq n$. The map φ is easily seen to be an isomorphism.

- Let G be a group and let $N \trianglelefteq G$. There is an obvious map from G to the quotient group G/N, namely the projection $\eta \colon G \to G/N$ defined by $\eta(x) = xN$ for $x \in G$. We see easily that this map η is an epimorphism. We shall refer to η as the *natural map* from G to G/N.

If $\varphi \colon G \to H$ is a homomorphism, then we define the *kernel* of φ to be the subset $\ker \varphi = \{g \in G \mid \varphi(g) = 1\}$ of G, and the *image* of φ to be the subset $\operatorname{im} \varphi = \{\varphi(g) \mid g \in G\}$ of H. We also use the notation $\varphi(G)$ for the image of φ, and $\varphi(K)$ for the set $\{\varphi(g) \mid g \in K\}$ for any $K \leqslant G$. For example, if $N \trianglelefteq G$ and $\eta \colon G \to G/N$ is the natural map, then we have $\ker \eta = N$ and $\eta(K) = KN/N$ for any $K \leqslant G$. (Observe that $\eta(K) = K/N$ if K contains N.)

PROPOSITION 11. *Let G and H be groups, and let $\varphi \colon G \to H$ be a homomorphism. Then $\ker \varphi \trianglelefteq G$, and $\varphi(K) \leqslant H$ for any $K \leqslant G$.* ∎

The following theorem is the cornerstone of group theory.

FUNDAMENTAL THEOREM ON HOMOMORPHISMS. *If G and H are groups and $\varphi \colon G \to H$ is a homomorphism, then there is an isomorphism $\psi \colon G/K \to \varphi(G)$ such that $\varphi = \psi \circ \eta$, where $K = \ker \varphi$ and $\eta \colon G \to G/K$ is the natural map; moreover, the map ψ is uniquely determined.*

(Many authors refer to this result as the "first isomorphism theorem;" these authors give appropriate renumbering to the other isomorphism theorems below.)

PROOF. If $xK = yK$ for some $x, y \in G$, then $y^{-1}x \in K$; this gives $1 = \varphi(y^{-1}x) = \varphi(y)^{-1}\varphi(x)$ and hence $\varphi(y) = \varphi(x)$. It is therefore possible to define a map $\psi\colon G/K \to \varphi(G)$ by letting $\psi(xK) = \varphi(x)$ for $xK \in G/K$. We leave it to the reader to verify that ψ has the indicated properties. ∎

As a consequence of the fundamental theorem, we see that any homomorphism $\varphi\colon G \to H$ can be regarded as the composition of an epimorphism (of G onto $\varphi(G)$) with a monomorphism (of $\varphi(G)$ into H).

The final three results of this section are also of primary importance.

FIRST ISOMORPHISM THEOREM. Let G be a group. If $N \trianglelefteq G$ and $H \leqslant G$, then $HN/N \cong H/H \cap N$.

(Note that $HN \leqslant G$ and $H \cap N \trianglelefteq H$ by Proposition 7, since $N \trianglelefteq G$.)

PROOF. Apply the fundamental theorem, taking φ to be the restriction to H of the natural map $\eta\colon G \to G/N$. ∎

The proof of the next result is straightforward, but somewhat tedious.

CORRESPONDENCE THEOREM. Let G and H be groups, and let $\varphi\colon G \to H$ be an epimorphism having kernel N. Then there is a bijective correspondence given by φ between the set of subgroups of G that contain N and the set of subgroups of H. If K is a subgroup of G containing N, then this correspondence sends K to $\varphi(K)$; if L is a subgroup of H, then the subgroup of G sent to L under this correspondence is $\varphi^{-1}(L) = \{x \in G \mid \varphi(x) \in L\}$. Moreover, if K_1 and K_2 are subgroups of G containing N, then:

- $K_2 \leqslant K_1$ iff $\varphi(K_2) \leqslant \varphi(K_1)$, and in this case we have $|K_1 : K_2| = |\varphi(K_1) : \varphi(K_2)|$.
- $K_2 \trianglelefteq K_1$ iff $\varphi(K_2) \trianglelefteq \varphi(K_1)$, and in this case the map from K_1/K_2 to $\varphi(K_1)/\varphi(K_2)$ sending xK_2 to $\varphi(x)\varphi(K_2)$ is an isomorphism. ∎

As a special case of the correspondence theorem, we have the following useful fact: If G is a group and $N \trianglelefteq G$, then every subgroup of G/N is of the form K/N for some subgroup K of G that contains N. (Here we take φ to be the natural map from G to G/N.)

SECOND ISOMORPHISM THEOREM. Let H and K be normal sub-groups of a group G. If H contains K, then $G/H \cong (G/K)/(H/K)$.

PROOF. Apply the correspondence theorem, taking φ to be the natural map from G to G/K. ∎

EXERCISES

1. Prove, or complete the sketched proof of, each result in this section.
2. We say that a group G has *exponent* e if e is the smallest positive integer such that $x^e = 1$ for every $x \in G$. Show that if G has exponent 2, then G is abelian. For what integers e is a group having exponent e necessarily abelian?
3. Let G be a finite group, and suppose that the map $\varphi: G \to G$ defined by $\varphi(x) = x^3$ for $x \in G$ is a homomorphism. Show that if 3 does not divide $|G|$, then G must be abelian. (See [2] for a generalization.)
4. Let g be an element of a group G, and suppose that $|G| = mn$ where m and n are coprime. Show that there are unique elements x and y of G such that $xy = g = yx$ and $x^m = 1 = y^n$. (In the case where m is a power of some prime p, we call x the *p-part* of g and y the *p'-part* of g; more generally, if π is a set of primes which includes all prime divisors of m and no prime divisors of n, then x and y are called the *π-part* and *π'-part*, respectively, of g.)
5. Let r, s, and t be positive integers greater than 1. Show that there is a finite group G having elements x and y such that x has order r, y has order s, and xy has order t.
6. Let X and Y be subsets of a group G. Are $<X> \cap <Y>$ and $<X \cap Y>$ necessarily equal? Are $<<X> \cup <Y>>$ and $<X \cup Y>$ necessarily equal?
7. Let G be a finite group and let $H \leqslant G$. Show that there is a subset T of G which is simultaneously a left transversal for H and a right transversal for H.
8. Suppose that $\mathcal{C}$ is a family of subsets of a group G which forms a partition of G, and suppose further that $g\mathcal{C} \in \mathcal{C}$ for any $g \in G$ and $C \in \mathcal{C}$. (Recall that a *partition* of a set S is a collection $\mathcal{S}$ of subsets of S with the property that every element of S lies in exactly one member of $\mathcal{S}$.) Show that $\mathcal{C}$ is the set of cosets of some subgroup of G.
9. Suppose that $\mathcal{C}$ is a family of subsets of a group G which forms a partition of G, and suppose further that $XY \in \mathcal{C}$ for any $X, Y \in \mathcal{C}$. Show that exactly one of the sets belonging to $\mathcal{C}$ is a subgroup of G, that this subgroup is normal in G, and that $\mathcal{C}$ consists of its cosets.

10. Prove the following generalization of Proposition 8: If G is a finite group and $H \leqslant G$ is such that $|G : H|$ is equal to the smallest prime divisor of $|G|$, then $H \trianglelefteq G$.

FURTHER EXERCISES

If $K \trianglelefteq H \leqslant G$, then H/K is called a *section* of G. We say that two sections H_1/K_1 and H_2/K_2 of G are *incident* if every coset of K_1 in H_1 has non-empty intersection with exactly one coset of K_2 in H_2, and vice versa. (In other words, two sections are incident if the relation of non-empty intersection gives a bijective correspondence between their elements.)

11. Show that incident sections are isomorphic.
12. (cont.) Suppose that $N \trianglelefteq G$ and $H \leqslant G$. Show that HN/N and $H/H \cap N$ are incident. (Exercises 11 and 12 provide an alternate proof of the first isomorphism theorem.)

If L/M is a section of G and $H \leqslant G$, then the *projection* of H on L/M is the subset of L/M consisting of those cosets of M in L which contain elements of H.

13. (cont.) Show that the projection of H on L/M is the subgroup $(L \cap H)M/M$ of L/M.

Let H_1/K_1 and H_2/K_2 be sections of a group G.

14. (cont.) Show that the projection of K_2 on H_1/K_1 is a normal subgroup of the projection of H_2 on H_1/K_1. The quotient group obtained thereby is called the projection of H_2/K_2 on H_1/K_1.
15. (cont.) Show that the projection of H_1/K_1 on H_2/K_2 and the projection of H_2/K_2 on H_1/K_1 are incident. Deduce the following result:

THIRD ISOMORPHISM THEOREM. Let $H_1, H_2 \leqslant G$, let $K_1 \trianglelefteq H_1$, and let $K_2 \trianglelefteq H_2$. Then

$$(H_1 \cap H_2)K_1/(H_1 \cap K_2)K_1 \cong (H_1 \cap H_2)K_2/(K_1 \cap H_2)K_2. \quad \blacksquare$$

(This result is also called the fourth isomorphism theorem, or Zassenhaus' lemma (after its discoverer, who proved it as a student at the age of 21), or even the butterfly lemma. This last name refers to the shape of the diagram showing the inclusion relations between the many subgroups involved in the statement of this result; such a diagram appears in [22, p. 62].)

2. Automorphisms

The set of automorphisms of a group G is denoted $\mathrm{Aut}(G)$. If φ and ρ are automorphisms of G, then their composition $\varphi \circ \rho$ is also an automorphism of G, and hence composition of mappings is a binary operation on $\mathrm{Aut}(G)$. This operation gives a group structure on $\mathrm{Aut}(G)$; the identity element is the trivial automorphism sending each element to itself, and the inverse of an automorphism φ is its inverse φ^{-1} as a set map. We call $\mathrm{Aut}(G)$ with this binary operation the *automorphism group* of G, and we may write $\varphi\rho$ in lieu of $\varphi \circ \rho$ for $\varphi, \rho \in \mathrm{Aut}(G)$.

Every element g of a group G defines a conjugation homomorphism $\varphi_g : G \to G$ by $\varphi_g(x) = gxg^{-1}$. (Observe that we indeed have $\varphi_g(xy) = \varphi_g(x)\varphi_g(y)$ and $\varphi_g(x^{-1}) = \varphi_g(x)^{-1}$.) Each such map φ_g is actually an automorphism of G, for given $x \in G$ we have $x = \varphi_g(g^{-1}xg)$, and if $\varphi_g(x) = \varphi_g(y)$ then we obtain $x = y$ by cancellation. These maps are called the *inner automorphisms* of G. We have $\varphi_g\varphi_h = \varphi_{gh}$ for any $g, h \in G$, since $g(hxh^{-1})g^{-1} = (gh)x(gh)^{-1}$ for any $x \in G$; consequently, there is a homomorphism from G to $\mathrm{Aut}(G)$ sending $g \in G$ to φ_g. The image of this homomorphism is called the *inner automorphism group* of G and is denoted $\mathrm{Inn}(G)$, while the kernel is called the *center* of G and is denoted $Z(G)$. Observe that

$$Z(G) = \{g \in G \mid \varphi_g(x) = x \text{ for all } x \in G\}$$
$$= \{g \in G \mid gx = xg \text{ for all } x \in G\},$$

and hence that $Z(G)$ consists of those elements of G which commute with every element of G. Clearly, G is abelian iff $Z(G) = G$.

If $\sigma \in \mathrm{Aut}(G)$ and $\varphi_g \in \mathrm{Inn}(G)$, then it is easily verified that $\sigma\varphi_g\sigma^{-1} = \varphi_{\sigma(g)}$. This shows that $\mathrm{Inn}(G) \trianglelefteq \mathrm{Aut}(G)$; the quotient group $\mathrm{Aut}(G)/\mathrm{Inn}(G)$ is called the *outer automorphism group* of G and is denoted $\mathrm{Out}(G)$. However, the term "outer automorphism" usually refers not to elements of $\mathrm{Out}(G)$ themselves, but rather to automorphisms of G which are not inner and which hence have non-trivial image in $\mathrm{Out}(G)$ under the natural map. If G is abelian, then all non-trivial automorphisms of G are outer in this sense, since in this case we have $\mathrm{Inn}(G) = 1$.

Given a group, we may wish to determine the structure of its automorphism group. This is often a difficult problem. We will now consider, in some detail, the automorphism groups of cyclic groups.

Let $G = <x> \cong \mathbf{Z}$, and let φ be an automorphism of G. Then $\varphi(x)$ must generate G; but the only generators of G are x and x^{-1}. Thus φ either fixes each element or sends each element to its inverse, and hence we have $\text{Aut}(G) \cong \mathbf{Z_2}$.

Now let $n \in \mathbb{N}$ and let $G = <x> \cong \mathbf{Z_n}$. Suppose that φ is an endomorphism of G. We have $\varphi(x) = x^m$ for some $0 \le m < n$; it follows that φ sends every element of G to its mth power. Hence we see that G has exactly n endomorphisms, namely the mth power maps σ_m for $0 \le m < n$.

PROPOSITION 1. Let $G = <x> \cong \mathbf{Z_n}$ for $n \in \mathbb{N}$, and for each $0 \le m < n$ let σ_m be the endomorphism of G sending x to x^m. Then $\text{Aut}(G)$ consists precisely of those σ_m for which $m \ne 0$ and $\gcd(m, n) = 1$. Furthermore, $\text{Aut}(G)$ is abelian and is isomorphic with the group $(\mathbb{Z}/n\mathbb{Z})^\times$ of units of the ring $\mathbb{Z}/n\mathbb{Z}$.

PROOF. The map σ_0 has trivial image and hence is not an automorphism. Now let $1 \le m < n$, and consider σ_m. If $\gcd(m, n) = 1$, then there exist integers a and b such that $am + bn = 1$, and hence $\sigma_m(x^a) = x^{am} = x^{1-bn} = x(x^n)^{-b} = x$, showing that σ_m is surjective. Since G is finite, a surjective map from G to G must also be injective; therefore $\sigma_m \in \text{Aut}(G)$. Conversely, if $\sigma_m \in \text{Aut}(G)$, then $x = \sigma_m(x^a) = x^{am}$ for some $a \in \mathbb{Z}$; since $x^{am-1} = 1$, we must have $am - 1 = bn$ for some $b \in \mathbb{Z}$, which forces $\gcd(m, n) = 1$. The first assertion now follows.

Given $1 \le m_1, m_2 < n$, we have $\sigma_{m_1}\sigma_{m_2} = \sigma_t = \sigma_{m_2}\sigma_{m_1}$, where $1 \le t < n$ is such that $m_1 m_2 \equiv t \pmod{n}$; therefore $\text{Aut}(G)$ is abelian. Since $(\mathbb{Z}/n\mathbb{Z})^\times = \{m + n\mathbb{Z} \mid 1 \le m < n, \gcd(m, n) = 1\}$, we easily see that the map sending σ_m to $m + n\mathbb{Z}$ is an isomorphism from $\text{Aut}(G)$ to $(\mathbb{Z}/n\mathbb{Z})^\times$. ∎

We define the *totient* of $n \in \mathbb{N}$ to be the number of positive integers that are both less than n and coprime to n. (This number is also referred to as the value at n of the Euler phi-function.) If we write $n = p_1^{a_1} \cdots p_r^{a_r}$ where the p_i are distinct primes, then the totient of n is $(p_1^{a_1} - p_1^{a_1-1}) \cdots (p_r^{a_r} - p_r^{a_r-1})$. We see immediately from Proposition 1 that the order of $\text{Aut}(\mathbf{Z_n})$ is the totient of n. In particular, $|\text{Aut}(\mathbf{Z_p})| = p - 1$ when p is prime.

PROPOSITION 2. Let p be a prime. Then $\text{Aut}(\mathbf{Z_p}) \cong \mathbf{Z_{p-1}}$.

PROOF. Let F be the field having p elements. By Proposition 1, $\text{Aut}(\mathbf{Z_p})$ is isomorphic with the multiplicative group $F^\times$ of non-zero elements of F. For each divisor d of $p-1$, let f_d be the number of elements of order d in $F^\times$, and let z_d be the number of elements of order d in $\mathbf{Z_{p-1}}$.

Let d be a divisor of $p-1$. If $x \in F^\times$ is an element whose order divides d, then x must be a root of $X^d - 1 \in F[X]$, which has at most d roots. Consequently, if x is of order d, then the powers of x are the only elements of $F^\times$ that are roots of $X^d - 1$, and therefore every element of $F^\times$ of order d must be contained in $<x> \cong \mathbf{Z_d}$. Hence either $f_d = 0$, or f_d is equal to the number of elements of order d in $\mathbf{Z_d}$.

Using Theorem 1.4, we see that if d is any divisor of $p-1$, then all elements of order d in $\mathbf{Z_{p-1}}$ are contained in a single cyclic subgroup of order d; therefore, z_d is equal to the number of elements of order d in $\mathbf{Z_d}$. The above paragraph now implies that $f_d \leq z_d$ for every $d \mid (p-1)$. But we have

$$\sum_{d \mid (p-1)} f_d = |F^\times| = p - 1 = |\mathbf{Z_{p-1}}| = \sum_{d \mid (p-1)} z_d,$$

which forces $f_d = z_d$ for every $d \mid (p-1)$. In particular, we have $f_{p-1} = z_{p-1} > 0$, and therefore $F^\times \cong \mathbf{Z_{p-1}}$. ∎

Let $G = <x> \cong \mathbf{Z_n}$ for $n \in \mathbb{N}$ and consider the mth power automorphism σ_m of G, where $1 \leq m \leq n$ and $\gcd(m, n) = 1$. A simple induction argument shows that $(\sigma_m)^k(x) = x^{m^k}$ for any $k \in \mathbb{N}$; thus the order of σ_m is the least positive integer k such that $x^{m^k} = x$, or equivalently the smallest $k \in \mathbb{N}$ such that $m^k \equiv 1 \pmod{n}$. If the order of σ_m is equal to the totient of n, then we say that m is a *primitive root modulo* n. (This terminology comes from classical number theory.) Clearly, $\text{Aut}(\mathbf{Z_n})$ is cyclic iff there exists a primitive root modulo n.

For composite n, the determination of the structure of $\text{Aut}(\mathbf{Z_n})$ lies more in the domain of number theory than group theory. The following result, which we shall not prove, characterizes those n for which $\text{Aut}(\mathbf{Z_n})$ is itself cyclic.

THEOREM 3. $\text{Aut}(\mathbf{Z_n})$ is cyclic iff $n = 2$ or 4, or $n = p^k$ or $2p^k$ for some odd prime p and some $k \in \mathbb{N}$. ∎

A proof of the equivalent result about the existence and non-existence of primitive roots modulo n is given in [9, Section 8.3].

Let φ be an automorphism of a group G, and let H be a subgroup of G. Then φ maps H isomorphically to a subgroup $\varphi(H)$ of G; we say that H is *fixed by* φ if $\varphi(H) = H$. In this case, the restriction of φ to H is an automorphism of H. If L is a subgroup of $\text{Aut}(G)$, then we say that H is *fixed by* L if H is fixed by every $\varphi \in L$. With this terminology, we see that H is normal in G iff H is fixed by $\text{Inn}(G)$. We say that H is a *characteristic subgroup* of G (or that H is *characteristic* in G) if H is fixed by $\text{Aut}(G)$. (Some authors denote this by H char G.) For example, the center $Z(G)$ is always a characteristic subgroup of G, for if $x \in Z(G)$ and $\varphi \in \text{Aut}(G)$, then we have $\varphi(x)y = \varphi(x\varphi^{-1}(y)) = \varphi(\varphi^{-1}(y)x) = y\varphi(x)$ for any $y \in G$, showing that $\varphi(x) \in Z(G)$ as required. It is clear that characteristic subgroups are normal, but the converse is not true. In fact, an infinite abelian group need not have any non-trivial proper characteristic subgroups; see Exercise 5.

We observed in Section 1 that being normal is not a transitive property of subgroups. However, being characteristic is transitive:

LEMMA 4. If K is a characteristic subgroup of H and H is a characteristic subgroup of G, then K is a characteristic subgroup of G.

PROOF. If $\varphi \in \text{Aut}(G)$, then the restriction of φ to H lies in $\text{Aut}(H)$ since H is characteristic in G, and hence the restriction of φ to K lies in $\text{Aut}(K)$ since K is characteristic in H. Therefore, any automorphism of G fixes K, as required. ∎

The reason that normality is not transitive stems from the fact that if $N \trianglelefteq G$, then the restriction to N of an element of $\text{Inn}(G)$ surely lies in $\text{Aut}(N)$ but need not lie in $\text{Inn}(N)$.

Recall that if x and y are elements of a group G, then the commutator of x and y is the element $[x,y] = xyx^{-1}y^{-1}$. We define the *derived group* of G to be the subgroup G' of G generated by the set of all commutators in G; that is, $G' = <\{[x,y] \mid x,y \in G\}>$. Clearly, G is abelian iff $G' = 1$; it is equally clear that if $H \leqslant G$, then $H' \leqslant G'$. It is important to remember that, in general, G' contains more than just the commutators of elements of G. Since for any elements x and y we have $[x,y]^{-1} = (xyx^{-1}y^{-1})^{-1} = yxy^{-1}x^{-1} = [y,x]$, we see via Proposition 1.2 that an arbitrary element of G' is a finite product of commutators of elements of G.

LEMMA 5. Let G be a group. Then G' is characteristic in G.

PROOF. Let $\varphi \in \operatorname{Aut}(G)$. We have $\varphi([x, y]) = [\varphi(x), \varphi(y)]$ for any $x, y \in G$. If $g \in G'$, then we have seen that g is a product of commutators; therefore the same is true for $\varphi(g)$, and hence $\varphi(g) \in G'$. Thus $\varphi(G') \leqslant G'$; but the same argument gives $\varphi^{-1}(G') \leqslant G'$ and hence $G' = \varphi(\varphi^{-1}(G')) \leqslant \varphi(G')$. Therefore $\varphi(G') = G'$, completing the proof. ∎

The derived group has the following important property:

PROPOSITION 6. Let G be a group, and let $N \trianglelefteq G$. Then G/N is abelian iff $G' \leqslant N$.

PROOF. For any $x, y \in G$, we have $[xG', yG'] = [x, y]G' = G'$; consequently, the derived group of G/G' is trivial, and so G/G' is abelian. Let $N \trianglelefteq G$. If $G' \leqslant N$, then by the second isomorphism theorem, G/N is isomorphic with a quotient of the abelian group G/G' and hence is abelian. Conversely, if G/N is abelian, then for any $x, y \in G$ we have $(xN)(yN) = (yN)(xN)$ and hence $[x, y] \in N$, which shows that $G' \leqslant N$. ∎

We shall complete this section with an important application of automorphism groups, namely the construction of semidirect products. Before that, we shall review the more familiar notion of direct products.

Let $n \in \mathbb{N}$, and let $G_1, \ldots, G_n$ be groups. We form their Cartesian product $G_1 \times \ldots \times G_n$, and we give this set a binary operation by defining $(g_1, \ldots, g_n)(g'_1, \ldots, g'_n) = (g_1 g'_1, \ldots, g_n g'_n)$. We call this operation "componentwise multiplication," and it gives the Cartesian product a group structure; the identity element is $(1, \ldots, 1)$, and the inverse of an arbitrary element $(g_1, \ldots, g_n)$ is $(g_1^{-1}, \ldots, g_n^{-1})$. We call $G_1 \times \ldots \times G_n$ with this binary operation the (*external*) *direct product* of $G_1, \ldots, G_n$. The order of the factors is irrelevant, for we easily see that $G_1 \times \ldots \times G_n \cong G_{\rho(1)} \times \ldots \times G_{\rho(n)}$ for any $\rho \in \Sigma_n$.

We observe that $G = G_1 \times \ldots \times G_n$ has the following properties:

- For each i, G has a normal subgroup H_i that is isomorphic with G_i; specifically, $H_i = \{(1, \ldots, g_i, \ldots, 1) \mid g_i \in G_i\}$

(where g_i appears in the ith place). Moreover, G/H_i is isomorphic with the direct product of the remaining G_j.

- Every $g \in G$ has a unique expression $g = h_1 \cdots h_n$ where $h_i \in H_i$; if $g = (g_1, \ldots, g_n)$, then $h_i = (1, \ldots, g_i, \ldots, 1)$ for each i (where again g_i appears in the ith place). Consequently, if the groups $G_1, \ldots, G_n$ are each finite, then we have $|G| = |G_1| \cdots |G_n|$.

Now suppose that G is a group having subgroups $H_1, \ldots, H_n$ such that the following conditions hold:

(1) $H_i \trianglelefteq G$ for each $1 \le i \le n$.
(2) Every $g \in G$ has a unique expression $g = h_1 \cdots h_n$ where $h_i \in H_i$ for each i.

Conditions (1) and (2) imply the following:

(3) $G = H_1 \cdots H_n$.
(4) $H_i \cap H_1 \cdots H_{i-1} H_{i+1} \cdots H_n = 1$ for each i.
(5) If $i \ne j$, then elements of H_i commute with elements of H_j.
(6) If $g = h_1 \cdots h_n$ and $g' = h_1' \cdots h_n'$, where $h_i, h_i' \in H_i$ for each i, then $gg' = (h_1 h_1') \cdots (h_n h_n')$.

Under these circumstances, we see that there is a unique isomorphism from G to the external direct product $H_1 \times \ldots \times H_n$, sending H_i to $1 \times \ldots \times H_i \times \ldots \times 1$. Consequently, we call G the (internal) direct product of its subgroups $H_1, \ldots, H_n$, and we may write $G = H_1 \times \ldots \times H_n$ (although this is a slight abuse of notation). It is important to note that if (1) holds, then (2) holds iff both (3) and (4) hold; hence to determine that a given group is a direct product, it suffices to verify either (1) and (2) or (1), (3), and (4). (Note that (4) reduces to $H_1 \cap H_2 = 1$ when $n = 2$.)

We now present some useful facts concerning direct products.

LEMMA 7. Let G be a group having normal subgroups H and K such that $G = HK$. Then $G/H \cap K = H/H \cap K \times K/H \cap K$.

PROOF. Note first that $L = H \cap K$ is normal in G by Proposition 1.7. We see from the correspondence theorem that H/L and K/L are normal subgroups of G/L, and clearly $(H/L) \cap (K/L)$ is trivial. Hence it remains only to show that $G/L = (H/L)(K/L)$. Let $g \in G$; then $g = hk$ for some $h \in H$ and $k \in K$ since $G = HK$, and thus $gL = hkL = hLkL \in (H/L)(K/L)$ as required. ∎

LEMMA 8. Let $n \in \mathbb{N}$, and write $n = p_1^{a_1} \cdots p_r^{a_r}$ where the p_i are distinct primes and the a_i are positive integers. Then we have $\mathbf{Z_n} \cong \mathbf{Z_{p_1^{a_1}}} \times \ldots \times \mathbf{Z_{p_r^{a_r}}}$.

PROOF. Let $P_i = <x_i> \cong \mathbf{Z_{p_i^{a_i}}}$ for each $1 \leq i \leq r$. We see easily that the order of $(x_1, \ldots, x_r) \in P_1 \times \ldots \times P_r$ is $p_1^{a_1} \cdots p_r^{a_r} = n$ and hence that $P_1 \times \ldots \times P_r \cong \mathbf{Z_n}$. ∎

This result has the following immediate consequence:

COROLLARY 9. If $\gcd(a, b) = 1$, then $\mathbf{Z_{ab}} \cong \mathbf{Z_a} \times \mathbf{Z_b}$. ∎

PROPOSITION 10. Suppose that a finite group G is the direct product of its subgroups $H_1, \ldots, H_n$, where the orders $|H_i|$ are pairwise coprime. Then any subgroup L of G is the direct product of $L \cap H_1, \ldots, L \cap H_n$.

PROOF. We consider the case $n = 2$, from which the general case follows easily by induction. Write $H = H_1$ and $K = H_2$, so that we have $G = H \times K$ and $\gcd(|H|, |K|) = 1$. Let $L \leqslant G$. Observe that we have $L \cap H \trianglelefteq L, L \cap K \trianglelefteq L$, and $(L \cap H) \cap (L \cap K) = 1$; therefore we can, inside L, construct the direct product $(L \cap H) \times (L \cap K)$. Every element g of L can be written as $g = hk$ for some $h \in H$ and $k \in K$, and to show that $L = (L \cap H) \times (L \cap K)$ it suffices to show that $h, k \in L$. Since h and k are commuting elements of coprime order, the order of hk equals the product of the orders of h and k. Corollary 9 now gives $<h> \times <k> \cong <hk>$. As we already have $<g> = <hk> \leqslant <h> \times <k>$, we now see that $h, k \in <g> \leqslant L$ as required. ∎

Let G be a group. Suppose that G has a subgroup H and a normal subgroup N such that $G = NH$ and $N \cap H = 1$; then we call G the (*internal*) *semidirect product of N by H*, and we write $G = N \rtimes H$. (This notation is common, but not standard; other possible notations include $N \ltimes H$ and $H \ltimes N$, and some authors do not adopt a notation.) If in addition we have $H \trianglelefteq G$, then G is the direct product of N and H. As an example, if we take $G = \Sigma_3$, $N = A_3$, and $H = <(1\ 2)>$, then we see easily that $G = N \rtimes H$; however, H is not normal in G, so G is not the direct product of N and H.

We now make some observations about semidirect products. Suppose throughout that $G = N \rtimes H$.

- We have $H = H/(N \cap H) \cong NH/N = G/N$ by the first isomorphism theorem. Consequently, if G is finite, then we have $|G| = |N||G : N| = |N||H|$.

- Since $G = NH$, each $x \in G$ can be written as $x = nh$ for some $n \in N$ and $h \in H$. Suppose that this could be done in two different ways; that is, suppose that $n_1 h_1 = n_2 h_2$ for some $n_1, n_2 \in N$ and $h_1, h_2 \in H$. Then we would have $n_2^{-1} n_1 = h_2 h_1^{-1} \in N \cap H = 1$, forcing $n_1 = n_2$ and $h_1 = h_2$. Hence each $x \in G$ has a unique expression $x = nh$ where $n \in N$ and $h \in H$.

- Let $x, y \in G$, and write $x = n_1 h_1$ and $y = n_2 h_2$ as above. We know that the element xy of G can be written as $n'h'$ for some unique $n' \in N$ and $h' \in H$; explicitly, we have $xy = n_1 h_1 n_2 h_1^{-1} \cdot h_1 h_2$, where $n' = n_1 h_1 n_2 h_1^{-1} \in N$ (since $N \trianglelefteq G$) and $h' = h_1 h_2 \in H$.

- Let $h \in H$. Since N is normal in G, conjugation by h maps N to N; consequently, we can define a map $\varphi_h \colon N \to N$ by $\varphi_h(n) = hnh^{-1}$ for $n \in N$. It is easy to show that φ_h is an automorphism of N, and also that $\varphi_h \circ \varphi_{h'} = \varphi_{hh'}$ for any $h' \in H$. Therefore, we have constructed a homomorphism $\varphi \colon H \to \mathrm{Aut}(N)$, where $\varphi(h) = \varphi_h$; we call φ the *conjugation homomorphism* of the semidirect product G. We now see that we have $(n_1 h_1)(n_2 h_2) = n_1 \varphi(h_1)(n_2) \cdot h_1 h_2$ for any $n_1, n_2 \in N$ and $h_1, h_2 \in H$, and thus the group operation of G can be expressed in terms of the group operations of N and H and the homomorphism φ.

- Suppose that the homomorphism $\varphi \colon H \to \mathrm{Aut}(N)$ defined above were the trivial homomorphism. Then we would have $nhn^{-1} = n\varphi(h)(n^{-1})h = nn^{-1}h = h$ for any $n \in N$ and $h \in H$, and consequently $H \trianglelefteq G$; therefore $G = N \times H$. Conversely, if $G = N \times H$, then elements of H commute with elements of N, and thus the homomorphism φ must be trivial.

- If the conjugation homomorphism $\varphi \colon H \to \mathrm{Aut}(N)$ is non-trivial, then the group G must be non-abelian, for there must be some $h \in H$ and $n \in N$ such that $hnh^{-1} = \varphi(h)(n) \neq n$, in which case h and n do not commute.

These observations suggest that if G is the internal semidirect product of N by H, then the behavior of G is governed by the structures of N and H and by the interaction between N and H inside G as determined by the conjugation homomorphism from H to $\mathrm{Aut}(N)$. Hence if we wish to develop a notion of an external semidirect product, it would seem prudent to take as our starting point two groups N and H along with a given homomorphism $\varphi\colon H \to \mathrm{Aut}(N)$, and then construct somehow a group that behaves as if it were an internal semidirect product $N \rtimes H$ having φ as its conjugation homomorphism.

With this in mind, let N and H be groups, and let φ be a given homomorphism from H to $\mathrm{Aut}(N)$. We define a binary operation on $N \times H$ by $(n_1, h_1)(n_2, h_2) = (n_1\varphi(h_1)(n_2), h_1 h_2)$. This definition gives $N \times H$ a group structure; the identity element is $(1,1)$ and the inverse of (n, h) is $(\varphi(h^{-1})(n^{-1}), h^{-1})$. We call this group the *(external) semidirect product of N by H corresponding to φ*, and we denote it by $G = N \rtimes_\varphi H$. (Again this notation is common but not standard; other common notations include $N \ltimes_\varphi H$ and $H_\varphi \times N$.) This group structure on the set $N \times H$ generally differs from the direct product group structure; in the direct product, elements of $1 \times H$ commute with elements of $N \times 1$, but that will not be the case here whenever φ is non-trivial.

The group $G = N \rtimes_\varphi H$ has subgroups $\mathcal{N} = N \times 1$ and $\mathcal{H} = 1 \times H$ that are isomorphic with N and H, respectively. For $(x, 1) \in \mathcal{N}$ and $(n, h) \in G$, we have

$$
\begin{aligned}
(n,h)(x,1)(n,h)^{-1} &= (n\varphi(h)(x), h)(\varphi(h^{-1})(n^{-1}), h^{-1}) \\
&= (n\varphi(h)(x)\varphi(h)(\varphi(h^{-1})(n^{-1})), hh^{-1}) \\
&= (n\varphi(h)(x)n^{-1}, 1) \in \mathcal{N},
\end{aligned}
$$

and hence $\mathcal{N} \trianglelefteq G$. Since we have $(n, h) = (n\varphi(1)(1), h) = (n, 1)(1, h)$ for any $(n, h) \in G$, we see that $G = \mathcal{N}\mathcal{H}$; since $\mathcal{N} \cap \mathcal{H}$ consists only of the identity element of G, we see that G is the internal semidirect product of $\mathcal{N}$ by $\mathcal{H}$. Furthermore, given $(n, 1) \in \mathcal{N}$ and $(1, h) \in \mathcal{H}$, we have $(1, h)(n, 1)(1, h)^{-1} = (\varphi(h)(n), 1)$, and hence the conjugation homomorphism from $\mathcal{H}$ to $\mathrm{Aut}(\mathcal{N})$ of $G = N \rtimes \mathcal{H}$ corresponds, after identifying $\mathcal{N}$ with N and $\mathcal{H}$ with H in the natural ways, with our original homomorphism $\varphi\colon H \to \mathrm{Aut}(N)$.

We conclude that given groups N and H and a homomorphism $\varphi\colon H \to \mathrm{Aut}(N)$, we can construct a new group, namely $N \rtimes_\varphi H$, which is the internal semidirect product of a subgroup isomorphic with N by a subgroup isomorphic with H. By identifying $\mathcal{N}$ with N and $\mathcal{H}$ with H, we can write $N \rtimes_\varphi H = \{nh \mid n \in N, h \in H\}$, where multiplication is defined by $(n_1 h_1)(n_2 h_2) = n_1 \varphi(h_1)(n_2) \cdot h_1 h_2$. Observe that in this instance we have $hnh^{-1} = \varphi(h)(n)$. As noted above, this group will be non-abelian whenever φ is a non-trivial homomorphism.

If φ and ψ are distinct homomorphisms from H to $\mathrm{Aut}(N)$, then the groups $N \rtimes_\varphi H$ and $N \rtimes_\psi H$ are by no means necessarily isomorphic. However, we are able to obtain a few results in this direction that will be useful in later sections.

PROPOSITION 11. Let H be a cyclic group and let N be an arbitrary group. If φ and ψ are monomorphisms from H to $\mathrm{Aut}(N)$ such that $\varphi(H) = \psi(H)$, then we have $N \rtimes_\varphi H \cong N \rtimes_\psi H$.

PROOF. Let $H = <x>$. Since $\varphi(H) = \psi(H)$ by hypothesis, we see that $\varphi(x)$ and $\psi(x)$ generate the same cyclic subgroup of $\mathrm{Aut}(N)$. Hence we can find $a, b \in \mathbb{Z}$ such that $\varphi(x)^a = \psi(x)$ and $\psi(x)^b = \varphi(x)$. As H is cyclic, we will have $\varphi(h^a) = \psi(h)$ and $\psi(h^b) = \varphi(h)$ for any $h \in H$. Now define $\tau \colon N \rtimes_\psi H \to N \rtimes_\varphi H$ by $\tau(nh) = nh^a$. Then

$$\begin{aligned}
\tau(n_1 h_1 n_2 h_2) &= \tau(n_1 \psi(h_1)(n_2) h_1 h_2) \\
&= n_1 \psi(h_1)(n_2)(h_1 h_2)^a \\
&= n_1 \varphi(h_1^a)(n_2) h_1^a h_2^a \\
&= n_1 h_1^a n_2 h_2^a = \tau(n_1 h_1) \tau(n_2 h_2),
\end{aligned}$$

which shows that τ is a homomorphism. We can similarly show that the map $\lambda \colon N \rtimes_\varphi H \to N \rtimes_\psi H$ defined by $\lambda(nh) = nh^b$ is also a homomorphism. To complete the proof, it suffices to show that the maps τ and λ are mutually inverse. The map $\tau \circ \lambda$ sends $nh \in N \rtimes_\varphi H$ to nh^{ab}. But $\varphi(x) = \psi(x)^b = (\varphi(x)^a)^b = \varphi(x^{ab})$, and φ is injective; therefore $x^{ab} = x$, and hence $h^{ab} = h$ for all $h \in H$. Thus $\tau \circ \lambda$ is the identity map on $N \rtimes_\varphi H$, and similarly $\lambda \circ \tau$ is the identity map on $N \rtimes_\psi H$, as required. ■

PROPOSITION 12. Let N and H be groups, let $\psi \colon H \to \mathrm{Aut}(N)$ be a homomorphism, and let $f \in \mathrm{Aut}(N)$. If $\hat{f}$ is the inner automorphism of $\mathrm{Aut}(N)$ induced by f, then $N \rtimes_{\hat{f} \circ \psi} H \cong N \rtimes_\psi H$.

PROOF. Define $\theta\colon N \rtimes_\psi H \to N \rtimes_{\hat{f}\circ\psi} H$ by $\theta(nh) = f(n)h$. We have

$$
\begin{aligned}
\theta(n_1 h_1 n_2 h_2) &= \theta(n_1 \psi(h_1)(n_2) h_1 h_2) \\
&= f(n_1) f(\psi(h_1)(n_2)) h_1 h_2 \\
&= f(n_1) \cdot (f \circ \psi(h_1) \circ f^{-1} \circ f)(n_2) \cdot h_1 h_2 \\
&= f(n_1) \cdot (\hat{f} \circ \psi)(h_1)(f(n_2)) \cdot h_1 h_2 \\
&= f(n_1) h_1 f(n_2) h_2 = \theta(n_1 h_1) \theta(n_2 h_2),
\end{aligned}
$$

which shows that θ is a homomorphism. But the homomorphism sending $nh \in N \rtimes_{\hat{f}\circ\psi} H$ to $f^{-1}(n)h \in N \rtimes_\psi H$ is inverse to θ, and therefore θ is an isomorphism. ■

As an example of a semidirect product, let $N \cong \mathbf{Z_n}$ for any $n \in \mathbb{N}$, let $H \cong \mathbf{Z_2}$, and let $\varphi\colon H \to \mathrm{Aut}(N)$ be the map that sends the generator h of H to the automorphism sending each element of N to its inverse (so that $\varphi(h) = \sigma_{n-1}$ in the notation given earlier in this section). The group $N \rtimes_\varphi H$ is called the *dihedral group* of order $2n$ and is denoted by D_{2n}. (Some authors denote this group by D_n.) It is non-abelian whenever $n > 2$; when $n = 2$, φ is the trivial homomorphism, and hence $D_4 \cong \mathbf{Z_2} \times \mathbf{Z_2}$. The group D_{2n} has a geometric interpretation as the symmetry group of a regular n-gon; the generator of N corresponds to rotation by $2\pi/n$ radians, and the generator of H corresponds to reflection through some fixed axis. There is also the *infinite dihedral group* $D_\infty = N \rtimes_\varphi H$, where $N \cong \mathbf{Z}$ and H and φ are as above.

We close this section with an illustration of how dihedral groups arise naturally in group theory.

PROPOSITION 13. Let s and t be elements of order 2 in a group G. (Such elements are called *involutions*.) Then $<s, t>$ is a dihedral group; in particular, $<s, t> = <st> \rtimes <s>$.

PROOF. Let $L = <s, t>$, $N = <st>$, and $H = <s>$. To show that $L = N \rtimes H$, we must show that $L = NH$, that $N \cap H = 1$, and that $N \trianglelefteq L$; to further show that L is dihedral, we must show that conjugation by s sends each element of N to its inverse. We have $s(st)s^{-1} = s^2 t s = t s = t^{-1} s^{-1} = (st)^{-1}$, and so this latter condition is satisfied. We similarly have $t(st)t^{-1} = (st)^{-1}$, and hence $N \trianglelefteq L$. Using Proposition 1.2, we find without difficulty that any element of L can be written as either $(st)^n$, $(st)^n s$, $(ts)^n$, or $(ts)^n t$ for some non-negative integer n. The latter two of these four forms can be reexpressed as $(st)^{-n}$ and $(st)^{-n-1}s$, respectively; it now follows that $L = NH$. Finally, if $s = (st)^n$ for some n, then we see easily that $s = (st)^{n-2}$, and from this we can conclude that $N \cap H = 1$. ∎

For a discussion of the importance of this particular result in the study of finite simple groups, see [7, Section 45].

Exercises

1. Let H be a subgroup of a cyclic group G. Show that every automorphism of H is the restriction to H of an automorphism of G.
2. Show that $\mathrm{Aut}(\mathbf{Z_8}) \cong \mathbf{Z_2} \times \mathbf{Z_2}$.
3. Show that $\mathrm{Aut}(\mathbf{Z_{p^2}}) \cong \mathbf{Z_{p^2-p}}$ for p a prime. (HINT: Let m be a primitive root modulo p, and show that either m or $m + p$ is a primitive root modulo p^2.)
4. Show that if $H \trianglelefteq G$, then any characteristic subgroup of H is normal in G.
5. Let F be a field, and consider F as a group under its additive operation. Show that F has no non-trivial proper characteristic subgroups.
6. Verify the claim made on page 19 that if (1) holds, then (2) holds iff (3) and (4) hold.
7. Verify the claim made on page 19 that (1) and (2) together imply (3) through (6).
8. Let G_1 and G_2 be groups, and let $H \leqslant G_1 \times G_2$. Define

$$P_1 = \{x \in G_1 \mid (x,y) \in H \text{ for some } y \in G_2\},$$
$$I_1 = \{x \in G_1 \mid (x,1) \in H\},$$

and analogously define subsets P_2 and I_2 of G_2.

   (a) Show for $i = 1,2$ that $P_i \leqslant G_i$ and $I_i \trianglelefteq P_i$.
   (b) Show that there is a unique isomorphism from P_1/I_1 to P_2/I_2 sending xI_1 to yI_2, where y is any element of G_2 such that $(x,y) \in H$.
   (c) Prove *Goursat's theorem*: There is a bijective correspondence between subgroups of $G_1 \times G_2$ and triples (S_1, S_2, φ), where S_i is a section of G_i $(i = 1,2)$ and $\varphi \colon S_1 \to S_2$ is an isomorphism. (Recall that a section of a group G is a group L/M, where $M \trianglelefteq L \leqslant G$.)

9. (cont.) Use Exercise 8 to give a different proof of Proposition 10.
10. Let $G = N \rtimes H$, and suppose that $N \leqslant K \leqslant G$. Show that $K = N \rtimes (K \cap H)$.

FURTHER EXERCISES

Let N and H be groups. An *extension of N by H* is a group E along with a monomorphism $i\colon N \to E$ and an epimorphism $\pi\colon E \to H$ such that $i(N) = \ker \pi$ (so that N imbeds in E as a normal subgroup, with the quotient group being isomorphic with H). We shall usually refer to an extension (E, i, π) simply by the group E; however, the nature of the maps i and π are important in distinguishing between extensions. We identify N with its image under i, and H with the quotient of E by N. As an example, let $\varphi\colon H \to \mathrm{Aut}(N)$ be a homomorphism; then the semidirect product $N \rtimes_\varphi H$ is an extension of N by H in an obvious way, taking i to be the inclusion map sending $n \in N$ to $(n, 1)$ and π to be the projection map sending (n, h) to h.

11. We say that an extension E of N by H is a *split extension* if there is a homomorphism $t\colon H \to E$ (called a *splitting map* for the extension) such that $\pi \circ t$ is the identity map on H, in which case $t(H)$ will be a transversal for N in E. Show that E is a split extension iff it is a semidirect product of N by H.

12. (cont.) Let Q be the quaternion group of order 8. (We can consider Q as the set $\{\pm 1, \pm i, \pm j, \pm k\}$ with multiplication given by the rules $i^2 = j^2 = k^2 = -1$ and $ij = k = -ji$.) Show that Q can be realized as a non-trivial extension in four ways—thrice as an extension of $\mathbf{Z_4}$ by $\mathbf{Z_2}$, and once as an extension of $\mathbf{Z_2}$ by $\mathbf{Z_2} \times \mathbf{Z_2}$—but that none of these extensions is split. (In other words, Q cannot be written non-trivially as a semidirect product.)

If E is an extension of N by H, then we cannot expect to find a homomorphism $t\colon H \to E$ such that $t(H)$ will be a transversal for N in E, for if such a t existed then E would be split. However, since $H \cong E/N$, we can always find a set map $t\colon H \to E$ whose image is a transversal for N; such a map is called a *section* of the extension. Moreover, we can always choose t so that $t(1) = 1$, in which case we say that t is *normalized*. (We use normalized sections instead of arbitrary sections in order to keep the notational complexity to a minimum.)

13. (cont.) Let t be a normalized section of an extension E. Let $\Psi\colon E \to \mathrm{Aut}(E)$ be the homomorphism sending an element of E to the corresponding inner automorphism of E. We shall, for $x \in E$, regard $\Psi(x)$ as being an automorphism of N, which is possible since $N \trianglelefteq E$. Define set maps $f\colon H \times H \to N$ and $\varphi\colon H \to \mathrm{Aut}(N)$ by

$$f(\alpha, \beta) = t(\alpha)t(\beta)t(\alpha\beta)^{-1},$$
$$\varphi(\alpha) = \Psi(t(\alpha)).$$

We call (f, φ) the *factor pair* arising from t. Show that (f, φ) has the following properties:

1 $f(\alpha, 1) = f(1, \alpha) = 1$ for every $\alpha \in H$, and $\varphi(1)$ is the identity in $\text{Aut}(N)$.

2 $\varphi(\alpha)\varphi(\beta) = \Psi(f(\alpha, \beta))\varphi(\alpha\beta)$ for $\alpha, \beta \in H$.

3 $f(\alpha, \beta)f(\alpha\beta, \gamma) = \varphi(\alpha)(f(\beta, \gamma))f(\alpha, \beta\gamma)$ for $\alpha, \beta, \gamma \in H$.

14. (cont.) Just as we were able to externalize the notion of semidirect product, so should we be able to externalize the notion of extension; that is, given groups N and H and appropriate additional data, we should be able to construct an extension of N by H. Using Exercise 13 as a guide, formulate such an external construction and prove that it works.

We shall return to these ideas in the further exercises to Section 9.

3. Group Actions

Let G be an arbitrary group. A (*left*) *action* of G on a set X is a map from $G \times X$ to X, with the image of (g, x) being denoted by gx, which satisfies the following conditions:

- $1x = x$ for every $x \in X$.
- $(g_1 g_2)x = g_1(g_2 x)$ for every $g_1, g_2 \in G$ and $x \in X$.

(Right actions are defined analogously and are used in lieu of left actions by many authors; however, in this book virtually all actions considered will be left actions.) If we have an action of G on X, then we say that G *acts* on X or that X is a *G-set*. If X is a G-set, then X is also an H-set for any $H \leqslant G$, as the action of G on X restricts to give an action of H on X.

For example, let $H \leqslant G$ and consider the coset space G/H. We have an obvious map from $G \times G/H$ to G/H, namely the left multiplication map sending (g, xH) to gxH. This is easily seen to be a left action of G on G/H. Whenever we refer to a coset space G/H as being a G-set, it is this action of G on G/H that we have in mind.

We now provide an alternate perspective on group actions.

PROPOSITION 1. There is a natural bijective correspondence between the set of actions of G on a set X and the set of homomorphisms from G to Σ_X.

PROOF. Let X be a G-set. For each $g \in G$, we define a map $\sigma_g \colon X \to X$ by $\sigma_g(x) = gx$ for $x \in X$. We see that $\sigma_{g^{-1}} \circ \sigma_g$ is the identity map on X, as $x = 1x = (g^{-1}g)x = g^{-1}(gx) = (\sigma_{g^{-1}} \circ \sigma_g)(x)$ for any $x \in X$; similarly, $\sigma_g \circ \sigma_{g^{-1}}$ is also the identity map. We conclude that each map σ_g has an inverse, namely $\sigma_{g^{-1}}$, and hence lies in Σ_X. Furthermore, the second condition in the definition of a group action ensures that we have $\sigma_{g_1 g_2} = \sigma_{g_1} \circ \sigma_{g_2}$ for any $g_1, g_2 \in G$. Consequently, we can define a homomorphism from G to Σ_X sending $g \in G$ to σ_g.

Conversely, suppose that $\sigma \colon G \to \Sigma_X$ is a homomorphism. We define a map from $G \times X$ to X by sending (g, x) to $\sigma(g)(x)$. One can easily check that this map is an action of G on X. We leave it to the reader to verify that these processes are inverse to one another, which establishes the desired bijective correspondence. ∎

If G has a proper subgroup H with $|G : H| = n$, then the action of G on G/H gives rise, via Proposition 1, to a non-trivial homomorphism from G to Σ_n. This fact is of particular use when G is assumed to be simple, for in this case such a homomorphism, being non-trivial, must be injective.

We say that the action of G on a set X is *faithful* (or that G acts *faithfully* on X) if the homomorphism from G to Σ_X corresponding to the action is injective. Equivalently, the action is faithful if the only element $g \in G$ satisfying $gx = x$ for every $x \in X$ is the identity element. If G acts faithfully on X, then we may refer to G as being a *permutation group* on X, since in this case G is imbedded isomorphically in Σ_X via its action on X.

CAYLEY'S THEOREM. G is isomorphic with a subgroup of Σ_G; in particular, if G is finite with $|G| = n$, then G is isomorphic with a subgroup of Σ_n.

PROOF. The group G acts on itself by left multiplication; this is the case $H = 1$ of the action of G on the coset space G/H discussed above. If $g \in G$ is such that $gx = x$ for all $x \in G$, then taking $x = g^{-1}$ we have $g^{-1} = gg^{-1} = 1$ and hence $g = 1$. Therefore, this action is faithful, and so we have a monomorphism from G to Σ_G; the result is now immediate. ∎

In theory, Cayley's theorem reduces the study of finite groups to the study of finite symmetric groups and their subgroups. While there have been occasions (as in [5]) where this philosophy has proven to be viable, in general the fact that finite groups are imbedded in symmetric groups has not influenced the methods used to study finite groups.

If X and Y are G-sets, then a function $\varphi \colon X \to Y$ is said to be a *G-set homomorphism* if it commutes with the actions of G, which is to say that $\varphi(gx) = g\varphi(x)$ for any $g \in G$ and $x \in X$. If φ is in addition bijective, then we say that φ is a *G-set isomorphism* and that X and Y are *isomorphic* G-sets, and we write $X \cong Y$ in this case.

Our present objective is to classify all G-sets up to isomorphism. In carrying this out, we will develop concepts that are of constant use in group theory.

Let X be a G-set. For each $x \in X$, we define the *orbit* of x to be the subset $Gx = \{gx \mid g \in G\}$ of X, and we define the *stabilizer* of x to be the subset $G_x = \{g \in G \mid gx = x\}$ of G. We easily see that Gx is itself a G-set under the action induced from that on X, and that G_x is a subgroup of G. A subset of X is a G-set under the action induced from X iff it is a union of orbits.

LEMMA 2. *If X is a G-set, then $G_{gx} = gG_xg^{-1}$ for any $g \in G$ and $x \in X$.*

PROOF. An element u of G stabilizes gx iff $g^{-1}ug$ stabilizes x, which occurs iff u lies in gG_xg^{-1}. ∎

We say that the action of G on X is *transitive* (or that G acts *transitively* on X) if there is some $x \in X$ such that $Gx = X$, or equivalently if for any $x_1, x_2 \in X$, there exists some $g \in G$ such that $gx_1 = x_2$. (Observe that if $Gx = X$ for some $x \in X$, then we must have $Gx = X$ for every $x \in X$.) A subset of X is a transitive G-set under the action induced from X iff it is comprised of a single orbit. For example, if $H \leqslant G$, then the action of G on G/H is transitive, since for $xH, yH \in G/H$ we have $(yx^{-1})xH = yH$.

PROPOSITION 3. *Any G-set has a unique partition consisting of transitive G-sets, namely its partition into orbits.*

PROOF. We first comment that if X has a partition consisting of transitive G-sets, then those sets must be the orbits, since the orbits are the only transitive subsets of X; this proves uniqueness. Since an arbitrary element $x \in X$ lies in the orbit Gx, to show existence it suffices to show that any two orbits of X under the action of G are either equal or disjoint. Let $x, y \in X$ and suppose that $Gx \cap Gy$ is non-empty. Then we have $g_1 x = g_2 y$ for some $g_1, g_2 \in G$. But then $y = g_2^{-1} g_1 x \in Gx$, and hence $Gy \subseteq Gx$; by symmetry, we have $Gx \subseteq Gy$ and hence $Gx = Gy$. ∎

We now see that in order to describe all G-sets, it suffices to describe all transitive G-sets. We have seen that coset spaces are examples of transitive G-sets; what we now show is that any transitive G-set is isomorphic with a coset space G/H for some $H \leqslant G$.

PROPOSITION 4. If X is a transitive G-set, then $X \cong G/G_x$ as G-sets for any $x \in X$.

PROOF. Let $x \in X$, and define $\varphi \colon G/G_x \to X$ by $\varphi(gG_x) = gx$ for $g \in G$. If $gG_x = g'G_x$ for $g, g' \in G$, then $g^{-1}g' \in G_x$ and hence $g^{-1}g'x = x$, giving $gx = g'x$; this shows that φ is a well-defined function, and by reversing the argument we see that φ is injective. We have $u\varphi(gG_x) = u(gx) = (ug)x = \varphi(ugG_x) = \varphi(u(gG_x))$ for any $u \in G$ and $gG_x \in G/G_x$, showing that φ is a G-set homomorphism. For any $y \in X$, by transitivity there exists some $g \in G$ such that $y = gx = \varphi(gG_x)$; this shows that φ is surjective. Therefore, φ is a G-set isomorphism, as required. ∎

Proposition 4 yields not only a classification of G-sets, but also the following useful result, often called the "orbit-stabilizer theorem:"

COROLLARY 5. Let X be a G-set. Then $Gx \cong G/G_x$ as G-sets for any $x \in X$; in particular, if G is finite, then $|Gx| = |G : G_x|$.

PROOF. This follows from Proposition 4 since Gx is a transitive G-set. ∎

Having shown that an arbitrary G-set is a union of transitive G-sets, and having determined all transitive G-sets up to isomorphism, we will have a good understanding of the structure of arbitrary G-sets once we answer the following question: When are two transitive G-sets isomorphic? We first require a lemma.

LEMMA 6. Let $\varphi \colon X \to Y$ be a homomorphism of G-sets, and let $x \in X$. Then $G_x \leqslant G_{\varphi(x)}$, and if φ is an isomorphism, then $G_x = G_{\varphi(x)}$.

PROOF. If $g \in G_x$, then $\varphi(x) = \varphi(gx) = g\varphi(x)$; it follows that $G_x \leqslant G_{\varphi(x)}$. If φ is an isomorphism, then by considering the G-set homomorphism $\varphi^{-1}: Y \to X$, we have $G_{\varphi(x)} \leqslant G_{\varphi^{-1}(\varphi(x))} = G_x$, and hence $G_x = G_{\varphi(x)}$. ■

PROPOSITION 7. If H and K are subgroups of G, then the G-sets G/H and G/K are isomorphic iff H and K are conjugate in G.

PROOF. Since G/H and G/K are transitive G-sets, we see from Lemma 2 that the set of stabilizers of the G-set G/H (resp., G/K) is precisely the set of conjugates of H (resp., K). If $G/H \cong G/K$ as G-sets, then it follows immediately from Lemma 6 that these sets of stabilizers are equal and in particular that H and K are conjugate. Conversely, suppose that $H = gKg^{-1}$ for some $g \in G$. Then H is the stabilizer of $gK \in G/K$, and so it follows from Proposition 4 that $G/K \cong G/H$ as G-sets. ■

Let X be a G-set. We say that X is *doubly transitive* (and that G acts *doubly transitively* on X) if whenever (x_1, x_2) and (y_1, y_2) are elements of $X \times X$ with $x_1 \neq x_2$ and $y_1 \neq y_2$, there exists some $g \in G$ such that $gx_1 = y_1$ and $gx_2 = y_2$. For example, the natural action of Σ_n on $\{1, \dots, n\}$ for $n \geq 2$ is doubly transitive. A doubly transitive G-set is clearly transitive. Some authors use the terminology "2-transitive" instead of doubly transitive since there is a more general notion of a k-transitive G-set for any $k \in \mathbb{N}$. (See [24, p. 250].)

A proper subgroup H of a group G is said to be *maximal* if there is no proper subgroup of G that properly contains H. For example, any subgroup of prime index is necessarily maximal by Theorem 1.6.

PROPOSITION 8. Let G be a group, let X be a doubly transitive G-set, and let $x \in X$. Then G_x is a maximal subgroup of G.

PROOF. By Proposition 4, we have $X \cong G/G_x$ as G-sets. Suppose that G_x is not maximal, so that $G_x < K < G$ for some subgroup K. Then there exist $g \in G$ and $k \in K$ such that $g \notin K$ and $k \notin G_x$. Since G/G_x is doubly transitive, there exists some $u \in G$ such that $uG_x = G_x$ and $u(kG_x) = gG_x$. This gives $u \in G_x$, and hence $uk \in K$. We also have $g^{-1}uk \in G_x$, and consequently $g \in K$. We have arrived at a contradiction; therefore, G_x is maximal. ■

We say that a non-empty subset B of a transitive G-set X is a *block* if B and $gB = \{gx \mid x \in B\}$ are either equal or disjoint for every $g \in G$. Observe that X is always a block and that any one-element subset of X is always a block. We say that the transitive G-set X is *primitive* if these are the only blocks.

PROPOSITION 9. There is a natural bijective correspondence between the set of blocks of a transitive G-set X which contain a given element x and the set of subgroups of G which contain G_x.

PROOF. Let B be a block containing $x \in X$, and consider the set $H_B = \{g \in G \mid gx \in B\}$; we wish to show that $H_B \leqslant G$. Clearly, $1 \in H_B$. Now let $g, g' \in H_B$. Since x and gx both lie in B, we see that $gB \cap B$ is non-empty and hence that $gB = B$. We now have $(gg')x = g(g'x) \in gB = B$ and hence $gg' \in H_B$. Also, for $g \in H_B$ we have $gx \in B$ and $g^{-1}(gx) = x \in B$; thus $g^{-1}B \cap B$ is non-empty, which forces $g^{-1}B = B$, from which we see that $g^{-1}x \in B$ and hence that $g^{-1} \in H_B$. Therefore, H_B is a subgroup of G; observe that $G_x \leqslant H_B$ since $x \in B$. Thus, to each block B of X containing x we can associate a subgroup H_B of G which contains G_x. We must show that this correspondence is bijective.

Let B and B' be distinct blocks of X which contain x. Then without loss of generality there exists some $y \in X$ such that $y \in B'$ and $y \notin B$; since G acts transitively on X, there exists some $g \in G$ such that $gx = y$. Now $g \in H_{B'}$ but $g \notin H_B$, and hence $H_B \neq H_{B'}$. This shows that the correspondence is injective.

Let H be a subgroup of G which contains G_x, and consider the subset $C = \{hx \mid h \in H\}$ of X. Clearly, C is non-empty; it is equally clear that if $g \in H$, then $gC = C$. Let $g \in G$ be such that $gC \cap C$ is non-empty. Then there exist $h_1, h_2 \in H$ such that $gh_1x = h_2x$; this gives $h_2^{-1}gh_1x = x$ and hence $h_2^{-1}gh_1 \in G_x \leqslant H$, and therefore $g \in H$. Consequently, if $g \in G$ is such that $gC \neq C$, then as we must have $g \notin H$, we see that $gC \cap C$ must be empty. Therefore C is a block. Now $H_C = \{g \in G \mid gx \in C\}$; clearly $H \leqslant H_C$. If $g \in H_C$, then $gx = hx$ for some $h \in H$; hence $h^{-1}gx = x$ and thus $h^{-1}g \in G_x \leqslant H$, giving $g \in H$. Therefore $H_C = H$, which shows that the correspondence is surjective. ∎

COROLLARY 10. Let X be a transitive G-set. Then X is primitive iff G_x is a maximal subgroup of G for every $x \in X$.

PROOF. Suppose X is primitive, and let $x \in X$. By Proposition 9, we see that the blocks of X which contain x correspond exactly to the subgroups of G which contain G_x. But by hypothesis there are only two such blocks, namely $\{x\}$ and X; as we already know of two subgroups of G containing G_x, namely G and G_x, we see that there is no proper subgroup of G which properly contains G_x. Therefore, G_x is maximal in G.

Conversely, suppose that every stabilizer is a maximal subgroup. Then by the same argument we see that an element $x \in X$ lies in exactly two blocks, namely $\{x\}$ and X, for if x were in some other block, then G_x would not be maximal. Consequently, X can have no other blocks besides itself and its one-element subsets, and so X is primitive. ∎

If X is a transitive G-set, then since all stabilizers are conjugate by Lemma 2, we see that if G_x is maximal for some $x \in X$, then G_x will be maximal for every $x \in X$; the statement of Corollary 10 can be modified accordingly.

COROLLARY 11. Any doubly transitive G-set is primitive.

PROOF. This follows from Proposition 8 and Corollary 10. ∎

In the remainder of this section, we give some elementary applications of the theory of group actions. As before, G denotes an arbitrary group.

PROPOSITION 12. If G is finite and $H, K \leqslant G$, then

$$|HK| = \frac{|H||K|}{|H \cap K|}.$$

PROOF. Let $X = G/K$; we consider X as an H-set under left multiplication. The orbit of $K \in G/K$ under the action of H is $\{hK \mid h \in H\} = HK$, and hence $|HK|$ is equal to $|K|$ multiplied by the number of cosets of K which lie in this orbit. The stabilizer H_K is easily seen to be $H \cap K$; therefore by Corollary 5, the orbit in question comprises $|H : H \cap K|$ cosets of K. ∎

For $x \in G$, we define the *centralizer* of x in G to be the set $C_G(x) = \{g \in G \mid gx = xg\}$ of elements in G that commute with x. We see easily that $C_G(x) \leqslant G$ for any $x \in G$. More generally, if $S \subseteq G$, then $C_G(S) = \{g \in G \mid gx = xg \text{ for all } x \in S\} = \cap_{x \in S} C_G(x)$

is called the centralizer of S in G. Observe that $Z(G) = C_G(G)$ and that $x \in Z(G)$ iff $C_G(x) = G$.

The *conjugacy class* of $x \in G$ is the set $\{gxg^{-1} \mid g \in G\}$ of all conjugates of x by elements of G. With this terminology, Proposition 1.10 asserts that the conjugacy class of an element ρ of Σ_n consists of all elements of Σ_n having the same cycle structure as ρ; the size of this conjugacy class can then be determined by an elementary combinatorial argument (see [24, p. 47]).

PROPOSITION 13. The conjugacy classes of G form a partition of G, and if G is finite then an element $x \in G$ has $|G : C_G(x)|$ conjugates in G.

PROOF. Let G act on itself by conjugation, so that $g \in G$ sends $x \in G$ to gxg^{-1}. (Verify that this is a left action.) The orbit of $x \in G$ under this action is $\{gxg^{-1} \mid g \in G\}$, which is the conjugacy class of x in G; therefore, the first assertion follows from Proposition 3. The second assertion follows from Corollary 5 and the observation that $G_x = C_G(x)$ for any $x \in G$. $\blacksquare$

A little thought will show that a subgroup of a group G is normal in G iff it comprises a union of conjugacy classes of G. We see from the proposition above that such a union is in fact disjoint. Consequently, the order of a normal subgroup of a finite group G must be a sum of orders of conjugacy classes of G.

For $H \leqslant G$, let $N_G(H) = \{g \in G \mid gHg^{-1} = H\}$; this set is called the *normalizer* of H in G. We see easily that $N_G(H) \leqslant G$ and that $H \trianglelefteq N_G(H)$; indeed, $N_G(H)$ is the largest subgroup of G in which H is normal, and so in particular we have $N_G(H) = G$ iff $H \trianglelefteq G$.

PROPOSITION 14. A subgroup H of a finite group G has exactly $|G : N_G(H)|$ conjugates in G. In particular, the number of conjugates of H in G divides $|G : H|$ and is equal to 1 iff $H \trianglelefteq G$.

PROOF. Let $\mathcal{P}(G)$ be the set of subsets of G, and let each $g \in G$ act on $\mathcal{P}(G)$ by sending $S \in \mathcal{P}(G)$ to gSg^{-1}. We easily see that this defines a left action of G on $\mathcal{P}(G)$. The orbit of $H \in \mathcal{P}(G)$ under this action is the set of conjugates of H in G, and the stabilizer of H is $N_G(H)$. The result now follows from Corollary 5. $\blacksquare$

Let G be a group and let H and K be subgroups of G. A *double coset* of H and K in G is a set $HxK = \{hxk \mid h \in H, \, k \in K\}$ for some $x \in G$. Suppose that $HxK \cap HyK$ is non-empty. Then there

exist $h, h' \in H$ and $k, k' \in K$ such that $hxk = h'yk'$; from this we see
that $x \in HyK$ and $y \in HxK$, and consequently that $HxK = HyK$.
Therefore, any two double cosets are either disjoint or equal, as is
the case with ordinary cosets.

Our final result generalizes Proposition 12.

PROPOSITION 15. If G is finite and $H, K \leqslant G$, then for any $x \in G$
we have

$$|HxK| = \frac{|H||K|}{|H \cap xKx^{-1}|} = \frac{|H||K|}{|x^{-1}Hx \cap K|}.$$

PROOF. As in the proof of Proposition 12, we consider G/K as
an H-set; then HxK is the union in G of those cosets of K which
lie in the orbit of xK in G/K, and consequently $|HxK|$ is equal
to $|K|$ multiplied by the number of cosets of K in that orbit. The
first equality follows from Corollary 5 once we observe via Lemma 2
that the stabilizer of xK under the action of H is $H \cap xKx^{-1}$. We
could prove the second equality by a similar argument in which we
consider the right action of K (by right multiplication) on the set
of right cosets of H in G. However, we can also prove the second
equality (and the first, for that matter) using Proposition 12, as
follows:

$$|HxK| = |HxKx^{-1}| = \frac{|H||xKx^{-1}|}{|H \cap xKx^{-1}|} = \frac{|H||K|}{|x^{-1}(H \cap xKx^{-1})x|}$$
$$= \frac{|H||K|}{|x^{-1}Hx \cap K|}.$$

(Here we use the fact that $g(S \cap T)g^{-1} = gSg^{-1} \cap gTg^{-1}$ for any
$g \in G$ and any $S, T \subseteq G$.) ■

EXERCISES

1. Show that a finite simple group whose order is at least $r!$ cannot
   have a proper subgroup of index r.
2. Show that a group G acts doubly transitively on a set X iff G_x acts
   transitively on $X - \{x\}$ for every $x \in X$. (Here we must assume
   that X has more than two elements.)
3. Show directly from the definitions (that is, without reference to
   Propositions 8 and 9) that a doubly transitive G-set is primitive.

4. Let G be the subgroup of Σ_5 generated by $(1\ 2\ 3\ 4\ 5)$, and let G act on $X = \{1,2,3,4,5\}$ in the canonical way. Show that this action is primitive, but not doubly transitive.

5. Let $N \trianglelefteq G$, and let $y \in N$. Show that the conjugacy class of y in G is a union of conjugacy classes of the group N. Show further that there is a bijective correspondence between the conjugacy classes of N which comprise the conjugacy class of y in G and the cosets of $NC_G(y)$ in G.

6. Let G be a finite group, and let r be the number of conjugacy classes of G. Show that $|\{(a,b) \in G \times G \mid ab = ba\}| = r|G|$.

7. Show that $C_G(gxg^{-1}) = gC_G(x)g^{-1}$ for any elements g and x of a group G.

8. Let $n \geq 5$, and assume that A_n is simple. (A proof of this fact is outlined in Exercise 7.8.) Use Exercise 1 to show that Σ_n has no proper subgroup of index less than n other than A_n.

FURTHER EXERCISES

Let X be a transitive G-set. A *system of imprimitivity* of X is a partition of X which is permuted by the action of G. Note that a system of imprimitivity of a G-set is itself a G-set.

9. Show that there is a bijective correspondence between the set of blocks which contain a given element of X and the set of systems of imprimitivity of X. (Observe that when X is finite, this implies that any two elements of X lie in the same number of blocks.)

10. Suppose that the G-set Y is an epimorphic image of a G-set X. Show that there is a G-set isomorphism between Y and some system of imprimitivity of X.

If X is a G-set, we use $[X]$ to denote the isomorphism class of X.

11. Let G be a finite group, and let $S(G)$ be the set of isomorphism classes of finite G-sets. Show that we can define sum and product operations on $S(G)$ by $[X] + [Y] = [X \cup Y]$ and $[X][Y] = [X \times Y]$.

12. (cont.) Let $B(G)$ be the set obtained from $S(G)$ by adjoining formal additive inverses of isomorphism classes, in the same way that $\mathbb{Z}$ is obtained from $\mathbb{N}$ by adjoining the additive inverses of positive integers. (The additive identity here will be the isomorphism class of the empty set.) Show that the operations defined above on $S(G)$ extend to give a commutative ring structure on $B(G)$. This ring $B(G)$ is called the *Burnside ring* of G.

13. (cont.) Show that any element of $B(G)$ can be written uniquely as a $\mathbb{Z}$-linear combination of isomorphism classes of transitive G-sets.

14. (cont.) Let $H \leqslant G$. Show that there is a unique ring homomorphism from $B(G)$ to $\mathbb{Z}$ which sends an isomorphism class $[X]$ to the number of elements of X fixed by H.

15. (cont.) Show that any non-zero homomorphism from $B(G)$ to $\mathbb{Z}$ arises from the construction in Exercise 14 and that the intersection of the kernels of all such homomorphisms is zero.

An element e of a ring R is called an *idempotent* if $e^2 = e$.

16. (cont.) Show that if G has no self-normalizing proper subgroup, then $B(G)$ has no idempotents other than the additive and multiplicative identities. (A subgroup H of G is called *self-normalizing* if $N_G(H) = H$. We shall see in Section 11 that there is an important class of groups, namely the nilpotent groups, that have no self-normalizing proper subgroups.)

2
The General Linear Group

This chapter presents an intensive look at an extremely important class of groups, the groups $GL(n, F)$ for F a field. While the material of this chapter does not play a large role in the remainder of this book, the ideas introduced here serve as an introduction to the manner in which group theory arises in modern mathematics. Section 4 defines the Borel and Weyl subgroups and establishes the Bruhat decomposition of $GL(n, F)$. Section 5 discusses unipotent and parabolic subgroups of $GL(n, F)$. In Section 6, we shift our attention to the groups $SL(n, F)$ and $PSL(n, F)$, culminating in a proof that $PSL(n, F)$ is simple except when $n = 2$ and $|F| \leq 3$.

4. Basic Structure

Let F be a field and let $n \in \mathbb{N}$. We denote by $\mathcal{M}_n(F)$ the set of all $n \times n$ matrices with entries in the field F. We often write such a matrix as $M = (m_{ij})$, where $m_{ij} \in F$ denotes the (i, j)-entry of M (the entry in the ith row and jth column). We define the *general linear group* $GL(n, F)$ to be the subset of $\mathcal{M}_n(F)$ consisting of all invertible matrices, or equivalently all matrices that have a non-zero determinant. $GL(n, F)$ forms a group under matrix multiplication; we denote the identity element by I.

More generally, given a finite-dimensional F-vector space V, we define the general linear group $GL(V)$ to be the group of all invertible linear transformations of V; here the group operation is composition of mappings. If we take $V = F^n$, then the resulting group is isomorphic in an obvious way with the matrix group $GL(n, F)$. Since any n-dimensional F-vector space is isomorphic with F^n, we lose nothing by restricting our attention to the groups $GL(n, F)$.

Recall that if F is a finite field, then F is determined up to isomorphism by its order $|F|$, which must be equal to p^a for some prime p and some $a \in \mathbb{N}$. (This result is due to E. H. Moore, founding head of the Department of Mathematics at the University of Chicago, who first announced it in 1893 at the first World Congress of Mathematicians in Chicago.) Consequently, if q is a prime power, then we can write $GL(n, q)$ in place of $GL(n, F)$, where F is the unique field of order q.

We start with an illustration of the importance of general linear groups in finite group theory.

PROPOSITION 1. Let E be a finite abelian group of exponent p, where p is prime. Then $\mathrm{Aut}(E) \cong GL(n, p)$, where $n \in \mathbb{N}$ is such that $|E| = p^n$.

(Recall that the exponent of a group is the least common multiple of the orders of its elements.)

PROOF. Let $F = \mathbb{Z}/p\mathbb{Z}$ be the field of p elements. We wish to give E the structure of an F-vector space. We define addition in E by $x + y = xy$. We define scalar multiplication for $\alpha \in F$ by $\alpha x = x^a$, where $a \in \mathbb{Z}$ is such that $\alpha = a + p\mathbb{Z}$; this is well-defined since E has exponent p. It is easy to verify that E now has an F-vector space structure; for example, we have $\alpha(x + y) = (xy)^a = x^a y^a = \alpha x + \alpha y$ for $\alpha \in F$ and $x, y \in E$. It now follows that any endomorphism of the group E is at the same time a linear transformation of the F-vector space E, and conversely; therefore, $\mathrm{Aut}(E) \cong GL(E) \cong GL(n, p)$, where $n = \dim_F E$. ∎

If E is as in the above proposition, and we let $\{x_1, \ldots, x_n\}$ be a basis for E as a vector space over the field of p elements, then it follows that as groups we have $E = <x_1> \times \ldots \times <x_n>$, where each group $<x_i>$ is cyclic of order p. We conclude that any finite abelian group of prime exponent p is isomorphic with a direct product of copies of $\mathbf{Z_p}$. Such groups are called *elementary abelian p-groups*. We define the *rank* of an elementary abelian p-group E to be n, where $|E| = p^n$. Observe that $\mathrm{Aut}(\mathbf{Z_p}) \cong \mathrm{GL}(1,p) \cong (\mathbb{Z}/p\mathbb{Z})^\times$ by Proposition 1; this was previously established in Proposition 2.1.

PROPOSITION 2. Let $n \in \mathbb{N}$, and let q be a prime power. Then

$$|\mathrm{GL}(n,q)| = \prod_{k=1}^{n}(q^n - q^{k-1}) = q^{\frac{n(n-1)}{2}}(q^n - 1)\cdots(q-1).$$

PROOF. To determine $|\mathrm{GL}(n,q)|$, it suffices to count the number of $n \times n$ matrices having entries in the field F of q elements and whose rows are linearly independent over F. To construct such a matrix, we can choose any non-zero vector in F^n as the first row; there are $q^n - 1$ such choices. For $1 < k \leq n$, the kth row can be any vector in F^n except for the q^{k-1} linear combinations of the previous $k - 1$ rows; hence there are $q^n - q^{k-1}$ choices for the kth row. The stated formula now follows. ∎

We now fix a field F and some $n \in \mathbb{N}$, and we write G instead of $\mathrm{GL}(n,F)$.

Let $M \in \mathcal{M}_n(F)$, and write $M = (m_{ij})$. The *main diagonal* of M consists of the entries m_{ii} for $1 \leq i \leq n$. We say that M is *diagonal* if its only non-zero entries appear on the main diagonal. We say that M is *upper triangular* if all entries of M lying below the main diagonal, namely those m_{ij} for which $i > j$, are zero.

PROPOSITION 3. The set B consisting of all invertible upper triangular matrices is a subgroup of G, called the *standard Borel subgroup*.

PROOF. It is easily verified that B is closed under matrix multiplication; hence it suffices to show that if $M \in B$, then the inverse N of M in G lies in B. Write $M = (m_{ij})$ and $N = (n_{ij})$. Since M is upper triangular, we see that the determinant of M is equal to the product of the entries on the main diagonal of M. As this determinant is non-zero, we must have $m_{ii} \neq 0$ for all i. Now $MN = I$, so $\sum_{k=1}^{n} m_{ik}n_{kj} = \delta_{ij}$ for any i and j. (Here δ_{ij} is the Kronecker delta symbol, which takes the value 1 if $i = j$, and 0 otherwise.) Taking $i = n$ shows that $n_{nj} = 0$ for all $j < n$, since $m_{nk} = 0$ for $k < n$ and $m_{nn} \neq 0$. If we now take $i = n - 1$, we find that $n_{(n-1)j} = 0$ for all

$j < n - 1$. Continuing this process, we find that $n_{ij} = 0$ whenever $i > j$, and hence that $N \in B$ as required. ∎

More generally, a *Borel subgroup* of G is any conjugate of the standard Borel subgroup B.

A *permutation matrix* is a matrix in which every row and column has a unique non-zero entry and all non-zero entries are equal to 1. For example, the identity matrix is a permutation matrix, and in fact every permutation matrix can be obtained from the identity matrix by switching columns (or rows). Every permutation matrix is orthogonal and thus has an inverse that is again a permutation matrix, namely its transpose. In particular, all $n \times n$ permutation matrices lie in G.

PROPOSITION 4. The set W consisting of all permutation matrices is a subgroup of G, called the *Weyl subgroup*.

PROOF. It suffices to show that the product of two permutation matrices is a permutation matrix. Let $M = (m_{ij})$ and $N = (n_{ij})$ be permutation matrices, and let $MN = P = (p_{ij})$. For any i and j, we see that $p_{ij} = 1$ if there exists k such that $m_{ik} = n_{kj} = 1$, and that $p_{ij} = 0$ otherwise. Given i, there is a unique k such that $m_{ik} = 1$, and there is a unique j such that $n_{kj} = 1$. Therefore, we see that $p_{ij} = 1$ for one and only one j; similarly, given j, we see that $p_{ij} = 1$ for exactly one i. Hence P is a permutation matrix. ∎

Let $V_n(F)$ denote the vector space of n-dimensional column vectors with entries in F, and let $\mathbf{v_1}, \ldots, \mathbf{v_n}$ be the standard basis. If we multiply $\mathbf{v_i}$ on the left by an $n \times n$ matrix M, then we obtain the ith column of M; we say that M *sends* $\mathbf{v_i}$ to the ith column of M.

PROPOSITION 5. $W \cong \Sigma_n$.

PROOF. Observe that any permutation matrix sends each $\mathbf{v_i}$ to some $\mathbf{v_k}$. Let $X = \{1, \ldots, n\}$. For each $w \in W$, we define a map $\varphi(w): X \to X$ by $\varphi(w)(i) = k$ for $1 \le i \le n$, where $1 \le k \le n$ is such that $\mathbf{v_k} = w\mathbf{v_i}$. If $\varphi(w)(i) = \varphi(w)(j)$ for some i and j, then the ith and jth columns of w must be equal; since w is a permutation matrix, this forces $i = j$. The map $\varphi(w)$ is thus injective, and hence bijective (since X is a finite set), for every $w \in W$; consequently, we have a map $\varphi: W \to \Sigma_n$. If $\varphi(w) = \varphi(w')$ for some $w, w' \in W$, then we have $w\mathbf{v_i} = w'\mathbf{v_i}$ for all $1 \le i \le n$; that is, the ith columns of w and w' are equal for every $1 \le i \le n$, which gives $w = w'$. Therefore,

φ is injective. If $\rho \in \Sigma_n$, then $\rho = \varphi(w)$, where w is the permutation matrix whose ith column is $\mathbf{v}_{\rho(i)}$ for every $1 \le i \le n$; therefore, φ is surjective. We leave it to the reader to verify that φ is a group homomorphism. ∎

We will often implicitly regard a permutation matrix w as being the element of Σ_n sending i to j, where $\mathbf{v_j}$ is the ith column of w. For example, the matrix

$$\begin{pmatrix} 0 & 0 & 1 & 0 \\ 0 & 0 & 0 & 1 \\ 1 & 0 & 0 & 0 \\ 0 & 1 & 0 & 0 \end{pmatrix}$$

corresponds to $(1\ 3)(2\ 4) \in \Sigma_4$, and so if w is this matrix then we may write $w(1) = 3$, and so forth. One observation about permutation matrices which will prove useful is that if $w(i) = j$, then for any $M \in \mathcal{M}_n(F)$, the jth row of wM is equal to the ith row of M, and the ith column of Mw is equal to the jth column of M.

Let $1 \le i, j \le n$ be distinct, and let $\alpha \in F$. We define $X_{ij}(\alpha)$ to be the $n \times n$ matrix whose (k, l)-entry is equal to α if $(k, l) = (i, j)$ and equal to δ_{kl} for all other (k, l). For example, $X_{23}(\alpha) \in \mathcal{M}_3(F)$ is the matrix

$$\begin{pmatrix} 1 & 0 & 0 \\ 0 & 1 & \alpha \\ 0 & 0 & 1 \end{pmatrix}.$$

These matrices $X_{ij}(\alpha)$, and their conjugates by elements of G, are called *transvections*. We leave to the reader the verification of the following properties of transvections:

LEMMA 6. Let $\alpha, \beta \in F$, and let i and j be distinct.

(i) $X_{ij}(\alpha)$ has determinant 1 and hence lies in G.

(ii) If $\alpha \ne 0$, then $X_{ij}(\alpha) \in B$ iff $i < j$.

(iii) $X_{ij}(\alpha)X_{ij}(\beta) = X_{ij}(\alpha + \beta)$, and hence $X_{ij}(\alpha)^{-1} = X_{ij}(-\alpha)$.

(iv) $[X_{ij}(\alpha), X_{jk}(\beta)] = X_{ik}(\alpha\beta)$ whenever i, j, k are distinct.

(v) If $w \in W$, then $wX_{ij}(\alpha)w^{-1} = X_{w(i)w(j)}(\alpha)$.

(vi) $X_{ij}(\alpha)$ sends $\mathbf{v_j}$ to $\mathbf{v_j} + \alpha\mathbf{v_i}$ and fixes $\mathbf{v_k}$ whenever $k \ne j$.

(vii) If $M \in \mathcal{M}_n(F)$, then the ith row of $X_{ij}(\alpha)M$ is equal to the sum of the ith row of M and α times the jth row of M, and for $k \ne i$ the kth row of $X_{ij}(\alpha)M$ is equal to the kth row of M. ∎

For distinct i and j, we define $X_{ij} = \{X_{ij}(\alpha) \mid \alpha \in F\}$; this is a subgroup of G by parts (i) and (iii) of Lemma 6. The subgroups X_{ij} are called *root subgroups* of G. (This terminology comes from the theory of Lie algebras.)

We now come to the main result of this section, in which we obtain the *Bruhat decomposition* of the group G. The following lemma contains the main thrust of the argument.

LEMMA 7. Let $M \in G$. Then there is a product b of upper triangular transvections such that the following property holds: For each $1 \leq i \leq n$, bM has exactly one row, say the k_ith row, whose entries in the first $i - 1$ columns are zero and which has a non-zero entry in the ith column.

PROOF. Let $M = (m_{ij}) \in G$. Since M is invertible, the first column of M must have some non-zero entry; let $1 \leq k_1 \leq n$ be such that $m_{k_1 1} \neq 0$ and $m_{i1} = 0$ for all $i > k_1$. For example, if we take $n = 5$ and $k_1 = 3$, then

$$M = \begin{pmatrix} * & * & * & * & * \\ * & * & * & * & * \\ * & * & * & * & * \\ 0 & * & * & * & * \\ 0 & * & * & * & * \end{pmatrix}$$

(where the symbol * denotes an arbitrary entry). We premultiply M by transvections of the form $X_{ik_1}(\alpha)$, where $i < k_1$, in such a way that the only non-zero entry in the first column of the resulting product $M' = (m'_{ij})$ lies in the k_1th row. Continuing with the above example, we have

$$M' = \begin{pmatrix} 0 & * & * & * & * \\ 0 & * & * & * & * \\ * & * & * & * & * \\ 0 & * & * & * & * \\ 0 & * & * & * & * \end{pmatrix}.$$

All of the transvections used to obtain M' from M lie in B, and hence $M'M^{-1} \in B$. As M' is again invertible, the second column of M' must have a non-zero entry in some row other than the k_1th row. Let $1 \leq k_2 \leq n$ be such that $k_2 \neq k_1$, $m'_{k_2 2} \neq 0$, and $m'_{i2} = 0$ for all $i > k_2$, $i \neq k_1$. Again we premultiply M' by transvections of the form $X_{ik_2}(\alpha)$, where $i < k_2$, in such a way that all entries in the second column of the resulting product M'' are zero except for those in rows k_1 and k_2. Once again, all of the transvections used to obtain M'' from M' lie in B, and so we have $M''M^{-1} \in B$. We continue this process, ultimately obtaining a matrix bM with the desired property, where $b \in B$ is a product of upper triangular transvections. For example, if we take $n = 5$ and $(k_1, k_2, k_3, k_4, k_5) = (3, 5, 4, 1, 2)$, then

$$bM = \begin{pmatrix} 0 & 0 & 0 & * & * \\ 0 & 0 & 0 & 0 & * \\ * & * & * & * & * \\ 0 & 0 & * & * & * \\ 0 & * & * & * & * \end{pmatrix}.$$

(Of course, the numbers $k_1, \ldots, k_n$ are just $1, \ldots, n$ under some reordering.) ∎

BRUHAT DECOMPOSITION THEOREM. $G = BWB$.

PROOF. Let $M \in G$, and define the numbers k_i and the matrix $b \in B$ as in the statement of Lemma 7. Let $w \in W$ be the permutation matrix whose k_ith column is v_i for each i. Then the ith row of wbM is equal to the k_ith row of bM for every i, and hence wbM is upper triangular. Thus $wbM \in B$, giving $M \in b^{-1}w^{-1}B \subseteq BWB$. ∎

We have now expressed G as a union of the double cosets BwB, as w ranges through W. We will now show that this union is disjoint. Again we first need a lemma.

LEMMA 8. If $w_1, w_2 \in W$ and $b \in B$ are such that $w_1bw_2 \in B$, then $w_2 = w_1^{-1}$.

PROOF. Let $1 \leq j \leq n$ be given, and let i be such that $w_1(i) = j$; then the jth row of w_1b is equal to the ith row of b. Let k be such that $w_2(k) = i$; then the kth column of w_1bw_2 is equal to the ith column of w_1b. Consider the (i, i)-entry β of b. We see that $\beta \neq 0$ since $b \in B$, and that β is the (j, i)-entry of w_1b and hence is also the (j, k)-entry of w_1bw_2. As $w_1bw_2 \in B$, this forces $j \leq k$. We now have a matrix $w_2^{-1}w_1^{-1} \in W$ which, for each $1 \leq j \leq n$, sends v_j to v_k, where $j \leq k \leq n$. The only permutation matrix having this property is the identity matrix; therefore $w_2^{-1}w_1^{-1} = I$, which proves the result. ∎

COROLLARY 9. If w and w' are distinct elements of W, then BwB and $Bw'B$ are disjoint; consequently, G is the disjoint union as w runs through W of the $n!$ double cosets BwB.

PROOF. Suppose that $w, w' \in W$ are such that $BwB \cap Bw'B$ is non-empty. Since two double cosets are either disjoint or equal, we have $BwB = Bw'B$. In particular, we have $w' = bwb'$ for some $b, b' \in B$, and hence $w^{-1}b^{-1}w' \in B$; by Lemma 8, this shows that w^{-1} is inverse to w', and hence that $w = w'$. This proves the first assertion, and the second now follows from the Bruhat decomposition theorem. ∎

The following somewhat technical observation, which we shall need in the next section, is an immediate consequence of what we have done.

COROLLARY 10. Let $M \in G$, and let w be the unique element of W such that $M \in BwB$. Then w sends $\mathbf{v_i}$ to $\mathbf{v_{k_i}}$ for each i, where the numbers k_i are as defined in the statement of Lemma 7. In particular, if M sends $\mathbf{v_1}$ to $\alpha_1\mathbf{v_1} + \ldots + \alpha_k\mathbf{v_k}$ where $\alpha_k \neq 0$, then w sends $\mathbf{v_1}$ to $\mathbf{v_k}$. ∎

We close this section with a result giving a smaller generating set for G than that given by the Bruhat decomposition. Once again, we start with a lemma.

LEMMA 11. Let $b \in B$. Then there exists a product t of transvections such that tb is a diagonal matrix having the same main diagonal entries as b.

PROOF. Let $b \in B$, and recall that the diagonal entries of b are non-zero. Here we adopt a procedure similar to that used in Lemma 7, except in order to preserve the entries along the main diagonal, we start at the last column instead of the first. We premultiply b by transvections $X_{in}(\alpha)$ so that the only non-zero entry of the nth column of the resulting matrix lies in the nth row. The nth diagonal entry of the resulting matrix is the same as that of b, and the $(n-1)$th diagonal entry is non-zero. We now premultiply this matrix by transvections $X_{i(n-1)}(\alpha)$ to obtain a matrix having a 2×2 diagonal block in the bottom right corner, with the diagonal entries in that block being equal to those in the corresponding block of b. By continuing this process, we obtain a diagonal matrix whose diagonal entries are the same as those of b. ∎

THEOREM 12. G is generated by the set consisting of all invertible diagonal matrices and all transvections.

PROOF. Since $G = BWB$ by the Bruhat decomposition theorem, it suffices to show that B and W are contained in the subgroup of G generated by the diagonal matrices and the transvections. It follows directly from Lemma 11 that B has this property.

As $W \cong \Sigma_n$ by Proposition 5, we see that W is generated by the permutation matrices that correspond to the transpositions; these are precisely the matrices obtained from the identity matrix by transposing two columns. Let i and j be distinct. We need to show that we can, by means of diagonal matrices and transvections, construct the matrix which sends $\mathbf{v_i}$ to $\mathbf{v_j}$, sends $\mathbf{v_j}$ to $\mathbf{v_i}$, and fixes every other $\mathbf{v_k}$. We find that the matrix $X_{ji}(1)X_{ij}(-1)X_{ji}(1)$ sends $\mathbf{v_i}$ to $\mathbf{v_j}$, sends $\mathbf{v_j}$ to $-\mathbf{v_i}$, and fixes all other $\mathbf{v_k}$. To obtain the permutation matrix that sends each of $\mathbf{v_i}$ and $\mathbf{v_j}$ to the other and fixes all other $\mathbf{v_k}$, we premultiply this matrix by the diagonal matrix whose (i, i)-entry is equal to -1 and whose other non-zero entries are equal to 1. This proves that W lies within the group generated by diagonal matrices and transvections. ∎

EXERCISES

1. Show that $GL(2, 2) \cong \Sigma_3$.
2. (cont.) Construct a monomorphism $\varphi \colon GL(1, 4) \to GL(2, 2)$ that corresponds to the inclusion of A_3 in Σ_3. (Recall that the field $\mathbb{F}_4$ of 4 elements can be written as $\{0, 1, \alpha, \alpha^2\}$, where $\alpha + 1 = \alpha^2$ and $\lambda + \lambda = 0$ for all $\lambda \in \mathbb{F}_4$.) Show further that the extension of φ to a map from $\mathbb{F}_4$ to $\mathcal{M}_2(\mathbb{F}_2)$, where $\mathbb{F}_2 = \{0, 1\}$ is the field of 2 elements, is a monomorphism of rings.
3. (cont.) Construct an explicit monomorphism from $GL(n, 4)$ to $GL(2n, 2)$ for any $n \in \mathbb{N}$.
4. (cont.) More generally, show for any $n \in \mathbb{N}$ and any prime power q that $GL(2n, q)$ has a subgroup that is isomorphic with $GL(n, q^2)$. Will $GL(n, q^m)$ always have a subgroup isomorphic with $GL(mn, q)$ for any $m, n \in \mathbb{N}$ and any prime power q?
5. Show that $GL(4, 2) \cong A_8$.
6. Let β be a non-trivial automorphism of the field F. Use β to construct an outer automorphism of $GL(n, F)$. Do all outer automorphisms of $GL(n, F)$ arise in this way?

FURTHER EXERCISES

7. A matrix is said to be *monomial* if each row and column has exactly one non-zero entry. Let N be the subset of $G = \mathrm{GL}(n, F)$ consisting of all monomial matrices. Show that $N \leqslant G$, that $T = B \cap N$ is the subgroup of G consisting of all diagonal matrices, that $N = N_G(T)$, and that $N = T \rtimes W$.

Let G be a group. Suppose that G has subgroups B and N of G satisfying the following conditions:

- G is generated by B and N.
- $T = B \cap N$ is a normal subgroup of N.
- $W = N/T$ is generated by a finite set S of involutions (elements of order 2). In addition, if for each $w \in W$ we choose some $\dot{w} \in N$ such that $\dot{w}T = w$, then we must have
  - $\dot{s}B\dot{w} \subseteq B\dot{w}B \cup B\dot{s}\dot{w}B$ for any $s \in S$ and $w \in W$,
  - $\dot{s}B\dot{s} \not\subseteq B$ for any $s \in S$.

In this case we say that B and N form a *BN-pair* of G, or that (G, B, N, S) is a *Tits system* (after Jacques Tits). We call B the *Borel subgroup* of G, and $W = N/B \cap N$ the *Weyl group* associated with the Tits system. The *rank* of the Tits system is defined to be $|S|$.

8. (cont.) Let $G = \mathrm{GL}(n, F)$, let B be the standard Borel subgroup of G, let N be the subgroup of G consisting of all monomial matrices, and let $T = B \cap N$. By Exercise 7 above, we know that we can regard $W = N/T$ as being imbedded in G as the group of permutation matrices. (By making this identification, we can replace $\dot{w}$ by w in the above formulas.) Let S be the subset of W consisting of those permutation matrices that are obtained from the identity matrix by switching two adjacent columns. (In other words, if we identify W with Σ_n as in Proposition 5, then S corresponds to the set $\{(1\,2), (2\,3), \dots, (n-1\,n)\}$.) Show that (G, B, N, S) is a Tits system of rank $n - 1$.

9. (cont.) Let G be a group with a BN-pair. Verify that, for $w \in W$, the set $B\dot{w}B$ is independent of the choice of $\dot{w} \in N$ such that $\dot{w}T = w$. Show that we have a Bruhat decomposition

$$G = \bigcup_{w \in W} BwB$$

in which the union is disjoint, where BwB is taken to mean $B\dot{w}B$ for any $\dot{w} \in N$ with $\dot{w}T = w$.

5. Parabolic Subgroups

In this section we again let $G = \mathrm{GL}(n, F)$ for some field F and some $n \in \mathbb{N}$, and we let $V_n(F)$ be the vector space of column vectors of length n having entries in F. We will often consider an element of G as being the matrix, with respect to the standard basis $\mathbf{v}_1, \dots, \mathbf{v}_n$, of an invertible linear transformation of $V_n(F)$.

A *complete flag* on $V_n(F)$ is a sequence of subspaces

$$0 \subset V_1 \subset V_2 \subset \dots \subset V_{n-1} \subset V_n = V_n(F).$$

We will use the notation $(V_1, \dots, V_n)$ for the above flag. As we use $\subset$ to denote proper containment, we must have $\dim_F V_i = i$ for each i. The *standard flag* is defined by $V_i = F\mathbf{v}_1 \oplus \dots \oplus F\mathbf{v}_i = V_{i-1} \oplus F\mathbf{v}_i$ for each i (where by convention $V_0 = 0$).

There is a natural action of G on the set of complete flags on $V_n(F)$; namely, if $(V_1, \dots, V_n)$ is a complete flag and $g \in G$, we define $g(V_1, \dots, V_n) = (gV_1, \dots, gV_n)$, where we view g as an invertible linear transformation of $V_n(F)$. (Each gV_i is a subspace of dimension i, and the gV_i retain the containment relations among the V_i, so that $(gV_1, \dots, gV_n)$ is again a complete flag.) It is easily seen that this definition gives a group action. Now let $(V_1, \dots, V_n)$ be the standard flag. We wish to show that every complete flag lies in the orbit of the standard flag and hence that the G-set of complete flags is transitive. If $(W_1, \dots, W_n)$ is a complete flag, then there are $\mathbf{w}_1, \dots, \mathbf{w}_n \in V_n(F)$ such that $\mathbf{w}_i \in W_i - W_{i-1}$ for each i (where again $W_0 = 0$). Let g be the matrix whose ith column is $\mathbf{w}_i$ for each i; then g is invertible, since $\{\mathbf{w}_1, \dots, \mathbf{w}_n\}$ is a basis for $V_n(F)$. As $g\mathbf{v}_i = \mathbf{w}_i$ for each i, we see that $(W_1, \dots, W_n) = g(V_1, \dots, V_n)$, proving our claim.

The stabilizer of a complete flag $(V_1, \dots, V_n)$ under this action is the set of $g \in G$ such that $(gV_1, \dots, gV_n) = (V_1, \dots, V_n)$, or equivalently such that $gV_i = V_i$ for each i. It is not hard to see that the stabilizer of the standard flag is exactly the standard Borel subgroup B of G. (This argument could be used to prove that B is a subgroup of G.) Since the G-set of complete flags is transitive, we now see via Lemma 3.2 that the Borel subgroups of G, being by definition the conjugates of B, are exactly the stabilizers of the complete flags on $V_n(F)$.

An upper triangular matrix is said to be *upper unitriangular* if all of its entries on the main diagonal are equal to 1, or equivalently if it is upper triangular and its only eigenvalue is 1. By imitating the proof of Proposition 4.3, we can show that the set U consisting of all invertible upper unitriangular matrices is a subgroup of B. If we let T denote the subset of G consisting of all diagonal matrices, then we see easily that $T \leqslant B$ and $U \cap T = 1$.

PROPOSITION 1. $B = U \rtimes T$.

PROOF. We need only show that $U \unlhd B$ and $B = UT$. Let $\varphi \colon B \to T$ be the map sending a matrix to the diagonal matrix having the same main diagonal. It is easily verified that φ is a homomorphism, and the kernel of φ is evidently U, showing that $U \unlhd B$. If $b \in B$, then since the restriction of φ to T is the identity map, we have $b\varphi(b)^{-1} \in \ker \varphi = U$, and hence $b \in U\varphi(b) \subseteq UT$; therefore $B = UT$. (Observe that if $n > 1$, then T is not normal in B, and hence B is not the direct product of U and T.) ∎

Consider the action of U and T on the standard flag $(V_1, \ldots, V_n)$. An element of U sends each $\mathbf{v_i}$ to the sum of $\mathbf{v_i}$ and some element of V_{i-1}; conversely, any matrix having this property lies in U. Therefore, U consists of the matrices that stabilize each V_i and that induce the identity transformation on each quotient V_i/V_{i-1}. On the other hand, T consists exactly of the matrices that stabilize each of $F\mathbf{v_1}, \ldots, F\mathbf{v_n}$; we say that T is the *common stabilizer* of the $F\mathbf{v_i}$.

An element of G is said to be *unipotent* if its characteristic polynomial is $(X - 1)^n$. We see, using Jordan form, that any unipotent element of G is conjugate to an element of U. We say that a subgroup of G is *unipotent* if all of its elements are unipotent. For example, U is a unipotent subgroup, and in fact U comprises all unipotent elements of B, since the roots of the characteristic polynomial of an upper triangular matrix are the entries on the main diagonal.

KOLCHIN'S THEOREM. Any unipotent subgroup of G is conjugate with a subgroup of U.

PROOF. Let H be a unipotent subgroup of G. Suppose that H stabilizes some complete flag on $V_n(F)$. Then H is contained in some Borel subgroup of G, and hence H is conjugate with a subgroup of B. Since all unipotent elements of B lie in U, we see that this subgroup of B with which H is conjugate is in fact a subgroup of U. Thus it suffices to show that H stabilizes some complete flag on $V_n(F)$.

We use induction on n; the case $n = 1$ is trivial. Suppose we can show that H stabilizes some one-dimensional subspace W of $V_n(F)$. Then H induces a group of unipotent transformations on the quotient $V_n(F)/W$. By induction, H stabilizes some complete flag on this quotient; by pulling this flag back to $V_n(F)$ and adding W, we obtain a complete flag on $V_n(F)$ which is stabilized by H. Therefore, it suffices to show that there is some $0 \neq \mathbf{v} \in V_n(F)$ such that $x\mathbf{v} = \mathbf{v}$ for all $x \in H$. This fact will be established in Proposition 13.28. ∎

We now move on to the more general situation. A *flag* on $V_n(F)$ is a nested sequence of non-zero subspaces of $V_n(F)$ which terminates in $V_n(F)$ and has no repeated terms; in other words, a flag is a sequence $(W_1, \ldots, W_r)$ of subspaces of $V_n(F)$, where

$$0 \subset W_1 \subset W_2 \subset \ldots \subset W_{r-1} \subset W_r = V_n(F).$$

Since $\subset$ denotes proper containment, we have $r \leq n$; a complete flag is simply a flag for which $r = n$. As before, the set of all flags on $V_n(F)$ is a G-set. However, it is not transitive. More precisely, two flags $(W_1, \ldots, W_r)$ and $(W_1', \ldots, W_s')$ lie in the same orbit iff $r = s$ and $\dim_F W_i = \dim_F W_i'$ for all i, a condition we summarize by saying that the flags have the same *dimension sequence*.

We say that a subgroup of G is a *parabolic subgroup* if it is the stabilizer of some flag on $V_n(F)$. Let $(W_1, \ldots, W_r)$ be a flag, and let P be the parabolic subgroup that is the stabilizer of this flag. Choose subspaces Y_i so that $W_i = W_{i-1} \oplus Y_i$ for each i. The *unipotent radical* of P is the subgroup U_P of P consisting of those matrices that induce the identity transformation on each W_i/W_{i-1}; for example, $U_B = U$. A *Levi complement* of U_P is the subgroup L_P of P that is the common stabilizer of the Y_i. We observe that a Levi complement is isomorphic with $\mathrm{GL}(y_1, F) \times \ldots \times \mathrm{GL}(y_r, F)$, where $y_i = \dim_F Y_i = \dim_F W_i - \dim_F W_{i-1}$ for each i. (In particular, any two Levi complements of U_P are isomorphic.) For example, if we take $Y_i = F\mathbf{v_i}$ for each i, we have $L_B = T$, which is clearly isomorphic with a direct product of n copies of $\mathrm{GL}(1, F) \cong F^\times$.

The following result, which the reader is asked to prove in the exercises, generalizes Proposition 1 to arbitrary parabolic subgroups.

PROPOSITION 2. If P is a parabolic subgroup of G, then we have $P = U_P \rtimes L_P$, where U_P is the unipotent radical of P and L_P is a Levi complement of U_P. Furthermore, $P = N_G(U_P)$. ∎

By a *subflag* of the standard flag $(V_1, \ldots, V_n)$, we mean a flag $(W_1, \ldots, W_r)$ in which each W_i is equal to some V_j; there are 2^{n-1} such subflags. A *staircase group* is a parabolic subgroup of G that is the stabilizer of some subflag of the standard flag.

The terminology "staircase group" arises from the appearance of the matrices in these groups. For example, the stabilizer of the subflag $0 \subset V_2 \subset V_3 \subset V_6$ of the standard flag on $V_6(F)$ consists of all matrices in $G = \mathrm{GL}(6, F)$ of the form

$$\begin{pmatrix} * & * & * & * & * & * \\ * & * & * & * & * & * \\ 0 & 0 & * & * & * & * \\ 0 & 0 & 0 & * & * & * \\ 0 & 0 & 0 & * & * & * \\ 0 & 0 & 0 & * & * & * \end{pmatrix}.$$

(Imagine a "staircase" separating the zero entries from the arbitrary entries.) The unipotent radical of this staircase group consists of all matrices in G of the form

$$\begin{pmatrix} 1 & 0 & * & * & * & * \\ 0 & 1 & * & * & * & * \\ 0 & 0 & 1 & * & * & * \\ 0 & 0 & 0 & 1 & 0 & 0 \\ 0 & 0 & 0 & 0 & 1 & 0 \\ 0 & 0 & 0 & 0 & 0 & 1 \end{pmatrix},$$

and the Levi complement of this staircase group corresponding to the canonical choices $Y_2 = F\mathbf{v_3}$ and $Y_3 = F\mathbf{v_4} \oplus F\mathbf{v_5} \oplus F\mathbf{v_6}$ consists of all matrices in G of the form

$$\begin{pmatrix} * & * & 0 & 0 & 0 & 0 \\ * & * & 0 & 0 & 0 & 0 \\ 0 & 0 & * & 0 & 0 & 0 \\ 0 & 0 & 0 & * & * & * \\ 0 & 0 & 0 & * & * & * \\ 0 & 0 & 0 & * & * & * \end{pmatrix}$$

and hence is isomorphic with $\mathrm{GL}(2, F) \times \mathrm{GL}(1, F) \times \mathrm{GL}(3, F)$. In general, the unipotent radical of any staircase group is contained in B.

From our earlier remarks, we see that the G-set of all flags is partitioned into the orbits of the subflags of the standard flag, since any flag has the same dimension sequence as, and hence lies in the orbit of, some subflag of the standard flag. Therefore, we see using Lemma 3.2 that the parabolic subgroups of G are exactly the conjugates of the staircase groups.

We will close this section by showing that there are exactly 2^{n-1} subgroups of G containing the subgroup B of upper triangular matrices, these subgroups being precisely the staircase groups. The following lemma will serve as our starting point in attempting to classify those subgroups containing B.

LEMMA 3. The only non-zero subspaces of $V_n(F)$ left invariant by B are the subspaces $V_1, \dots, V_n$ that appear in the standard flag.

PROOF. We first observe that B fixes each of the spaces V_i. Now let V be a non-zero subspace of $V_n(F)$ that is stabilized by B. As we clearly have $V \subseteq V_i$ for some i, there is a minimal $1 \le k \le n$ such that $V \subseteq V_k$ and $V \not\subseteq V_{k-1}$. By the minimality of k, V contains an element of the form $\sum_{j=1}^{k} \alpha_j \mathbf{v_j}$, where the α_j lie in F and $\alpha_k \ne 0$. This element is sent to $\mathbf{v_k}$ by b^{-1}, where

$$
b = \begin{pmatrix}
1 & 0 & \cdots & \alpha_1 & \cdots & 0 \\
 & 1 & \cdots & \alpha_2 & \cdots & 0 \\
 & & \ddots & \vdots & & \vdots \\
 & & & \alpha_k & \cdots & 0 \\
 & & & & \ddots & \vdots \\
 & & & & & 1
\end{pmatrix}
$$

Since $b^{-1} \in B$ and B stabilizes V, this shows that $\mathbf{v_k} \in V$.

Now let $\sum_{j=1}^{k} \alpha_j \mathbf{v_j}$ be an arbitrary element of $V_k - V_{k-1}$. This element is the image of $\mathbf{v_k}$ under the matrix b defined as above and hence lies in V. Thus V contains $V_k - V_{k-1}$, and hence V_k. We conclude that $V = V_k$. ∎

THEOREM 4. The only subgroups of G that contain B are the staircase groups.

PROOF. Suppose that $B \le H \le G$. By Lemma 3, the only subspaces of $V_n(F)$ that H could leave invariant are those that comprise the standard flag $(V_1, \dots, V_n)$. Suppose that the subspaces of $V_n(F)$ left invariant by H are $V_{a_1}, \dots, V_{a_r}$, where $1 \le a_1 < \dots < a_r = n$. We wish to show that H is in fact equal to the staircase group that is the stabilizer of the flag $(V_{a_1}, \dots, V_{a_r})$. Instead of giving a formal proof of the general case, we will prove the special cases $r = 1$ and $r = 2$; after this, we will briefly discuss the general case, which is not any more conceptually difficult than the special cases but which can be notationally unwieldy.

Suppose that the only subspaces of $V_n(F)$ that are left invariant by H are 0 and $V_n(F)$; we are to prove that $H = G$. The subspace of $V_n(F)$ spanned by $\{h\mathbf{v_1} \mid h \in H\}$ is left invariant by H and hence must equal all of $V_n(F)$. In particular, there is some $h \in H$ such that $h\mathbf{v_1} = \alpha_1\mathbf{v_1} + \ldots + \alpha_n\mathbf{v_n}$, where $\alpha_n \neq 0$. We know from the Bruhat decomposition theorem that $h \in BwB$ for some $w \in W$, and by Corollary 4.10 we see that w sends $\mathbf{v_1}$ to $\mathbf{v_n}$. Since $B \leqslant H$, this element w lies in H. Now let $1 < j \leq n$ be such that w sends $\mathbf{v_j}$ to $\mathbf{v_1}$. Then $X_{1j} \leqslant B \leqslant H$, and since $w(1) = n$ and $w(j) = 1$, we see from part (v) of Lemma 4.6 that $X_{n1} = wX_{1j}w^{-1} \leqslant H$. We now wish to show that each root subgroup X_{ij} lies in H; since the diagonal matrices lie in B and hence in H, it would then follow from Theorem 4.12 that $H = G$. We know already that $X_{n1} \leqslant H$ and that if $i < j$ then $X_{ij} \leqslant B \leqslant H$. Now for any $\alpha \in F$ and any distinct $1 < i, j < n$, we see by using part (iv) of Lemma 4.6 that $X_{nj}(\alpha) = [X_{n1}(\alpha), X_{1j}(1)] \in H$; $X_{i1}(\alpha) = [X_{in}(\alpha), X_{n1}(1)] \in H$; and $X_{ij}(\alpha) = [X_{i1}(\alpha), X_{1j}(1)] \in H$. Therefore, $X_{ij} \in H$ for all i, j, as required.

Now suppose that H leaves invariant not only 0 and $V_n(F)$, but also exactly one other subspace, namely V_m for some $1 \leq m < n$. Let P be the staircase group that is the stabilizer of the flag $0 \subset V_m \subset V_n$; we know that $H \leqslant P$, and we wish to show that $H = P$. By Proposition 2, P is the semidirect product of its unipotent radical U_P by a Levi complement L_P, and we have already observed that $U_P \leqslant B$ since P is a staircase group. Hence it suffices to show that $L_P \leqslant H$. We can choose L_P to be a direct product of subgroups $K_1 \cong \mathrm{GL}(m, F)$ and $K_2 \cong \mathrm{GL}(n - m, F)$, where elements of K_1 (resp., K_2) have their only non-zero entries in the first m (resp., last $n - m$) rows and columns. Using Theorem 4.12, we see that K_1 (resp., K_2) is generated by the diagonal matrices and the root subgroups X_{ij} for distinct $1 \leq i, j \leq m$ (resp., $m + 1 \leq i, j \leq n$). The diagonal matrices lie in B and hence in H, so in order to show that $L_P \leqslant H$, it suffices to show that these particular root subgroups lie in H. Since H stabilizes V_m and no non-zero proper subspace thereof, we see by considering the subspace spanned by $\{h\mathbf{v_1} \mid h \in H\}$ that there must be some $h \in H$ sending $\mathbf{v_1}$ to $\alpha_1\mathbf{v_1} + \ldots + \alpha_m\mathbf{v_m}$, where $\alpha_m \neq 0$. If $w \in W$ is such that $h \in BwB$, then w sends $\mathbf{v_1}$ to $\mathbf{v_m}$ by Corollary 4.10, and as in the previous paragraph we can conclude from this that $X_{m1} \in H$. Calculations with commutators can now be used to show that $X_{ij} \leqslant H$ for all distinct $1 \leq i, j \leq m$. Similarly, by considering the subspace spanned by $\{h\mathbf{v_{m+1}} \mid h \in H\}$, we see that there is some $h' \in H$ that sends $\mathbf{v_{m+1}}$ to $\beta_1\mathbf{v_1} + \ldots + \beta_n\mathbf{v_n}$, where $\beta_n \neq 0$; Corollary 4.10 can now be used to show that if $w' \in W$ is such that $h' \in Bw'B$, then w' sends $\mathbf{v_{m+1}}$ to $\mathbf{v_n}$. (Since h stabilizes V_m, in the notation of Corollary 4.10 we have $\{k_1, \ldots, k_m\} = \{1, \ldots, m\}$, and some thought will show that we must then have $k_{m+1} = n$.) The same argument as before now shows that $X_{n(m+1)} \leqslant H$, from which we can conclude via commutator calculations that $X_{ij} \leqslant H$ for all distinct $m + 1 \leq i, j \leq n$. Therefore $H = P$.

In the general case where H leaves invariant the flag $(V_{a_1}, \ldots, V_{a_r})$ whose stabilizer is the staircase group P, to show that $H = P$ it suffices to show that a Levi complement L_P lies in H. The group L_P is a direct product of r general linear groups, and so by Theorem 4.12 it suffices to show that H contains certain root subgroups; by appropriate choice of L_P, the subgroups in question become all X_{ij} for distinct $a_{k-1} < i, j \leq a_k$ for some k (taking $a_0 = 0$). We accomplish this by first showing that each $X_{a_k(a_{k-1}+1)} \in H$, and then by using commutator arguments as above. ∎

EXERCISES

Let $G = \mathrm{GL}(n, F)$, where $n \geq 2$.

1. Suppose that $g \in G$ fixes some $0 \neq \mathbf{v} \in V_n(F)$ and induces the identity transformation on $V_n(F)/F\mathbf{v}$. Show that g is conjugate with an element of the root subgroup X_{12}.

2. We say a basis $\mathcal{B}$ of $V_n(F)$ *belongs* to a complete flag $(V_1, \ldots, V_n)$ if each V_i contains exactly one element of $\mathcal{B}$ that is not in V_{i-1}.

   (a) Show that if $(V_1, \ldots, V_n)$ and $(W_1, \ldots, W_n)$ are complete flags on $V_n(F)$, then there is a basis of $V_n(F)$ that belongs to both flags.

   (b) Use part (a) to give a different proof of the Bruhat decomposition theorem.

3. Let P and Q be distinct parabolic subgroups of G containing the standard Borel subgroup B of G (so that, by Theorem 4, P and Q are staircase groups).

   (a) Show that P and Q are not conjugate in G.
   (b) If L_P and L_Q are corresponding Levi complements, find conditions on P and Q that determine when L_P and L_Q will be conjugate in G.

4. Prove Proposition 2.

5. Prove the following generalization of Lemma 3: If P is a parabolic subgroup of G which is the stabilizer of the flag $(W_1, \ldots, W_r)$, then the W_i are the only subspaces of $V_n(F)$ left invariant by P.

6. (cont.) Using Exercise 5, show that any parabolic subgroup of G is self-normalizing.

7. Complete the outlined proof of Theorem 4.

8. Assume that the case $r = 1$ of Theorem 4 holds; we sketch an alternate proof of the general case of Theorem 4. We use induction on n, where $G = \mathrm{GL}(n, F)$. Let H be a subgroup of G which contains B, and suppose that H stabilizes V_m, where $1 \le m < n$. Let S be the stabilizer of V_m; observe that $H \le S$. We can write $S = U \rtimes (K_1 \times K_2)$, where the elements of $K_1 \cong \mathrm{GL}(m, F)$ have their only non-zero entries in the first m rows and columns, and where K_2 is a direct product of general linear groups whose elements have their non-zero entries concentrated in the last $n - m$ rows and columns. Since $U \le B \le H$, by Exercise 2.10 we have $H = U \rtimes Q$ for some $Q \le K_1 \times K_2$. Use Goursat's theorem (Exercise 2.8), Exercise 6, and the induction hypothesis to analyze Q and thereby conclude that H is a staircase group.

6. The Special Linear Group

The *special linear group* is the subgroup $\mathrm{SL}(n, F)$ of $\mathrm{GL}(n, F)$ consisting of all matrices having determinant 1. In other words, $\mathrm{SL}(n, F)$ is the kernel of the homomorphism $\det \colon \mathrm{GL}(n, F) \to F^\times$, and hence $\mathrm{SL}(n, F) \trianglelefteq \mathrm{GL}(n, F)$.

PROPOSITION 1. Let $n \in \mathbb{N}$, and let q be a prime power. Then

$$|\mathrm{SL}(n, q)| = \prod_{k=1}^{n-1} (q^{n+1} - q^k) = q^{\frac{n(n-1)}{2}} (q^n - 1) \cdots (q^2 - 1).$$

PROOF. The homomorphism det is clearly surjective, and so we have $F^\times \cong \mathrm{GL}(n, F)/\mathrm{SL}(n, F)$ for any field F by the fundamental theorem on homomorphisms. In particular, if $|F| = q$, then we have $|\mathrm{GL}(n, q) : \mathrm{SL}(n, q)| = q - 1$; the result now follows from Proposition 4.2. ∎

We showed in Theorem 4.10 that $\mathrm{GL}(n, F)$ is generated by the root subgroups and the diagonal subgroup. We now establish the analogous result for $\mathrm{SL}(n, F)$.

PROPOSITION 2. $\mathrm{SL}(n, F)$ is generated by the root subgroups X_{ij}.

PROOF. We first remark that by part (i) of Lemma 4.6, every transvection lies in $\mathrm{SL}(n, F)$. The result will follow from the following statements:

(1) An element of $\mathrm{SL}(n, F)$ can be premultiplied by transvections to obtain an upper triangular matrix.
(2) An upper triangular element of $\mathrm{SL}(n, F)$ can be premultiplied by transvections to obtain an upper unitriangular matrix.
(3) An upper unitriangular matrix can be premultiplied by transvections to obtain the identity matrix.

We will sketch (1) and (2); (3) follows immediately from Lemma 4.11.

By Lemma 4.7, we can premultiply any element of $\mathrm{SL}(n, F)$ by transvections to obtain a matrix with the following property: For each i, there is exactly one row of the matrix whose entries in the first $i - 1$ columns are zero and that has a non-zero entry in the ith column. We can premultiply this matrix by matrices of the form $X_{ji}(1)X_{ij}(-1)X_{ji}(1)$ to obtain an upper triangular matrix. This proves (1).

Let $a, b, c \in F$, with $ac \neq 0$. We observe that the matrix $\left(\begin{smallmatrix} a & b \\ 0 & c \end{smallmatrix}\right)$, when premultiplied by $X_{21}(-1)X_{12}(1 - c^{-1})X_{21}(c)$, gives the matrix $\left(\begin{smallmatrix} ac & c+bc-1 \\ 0 & 1 \end{smallmatrix}\right)$. Now consider an upper triangular matrix in $\mathrm{SL}(n, F)$ having $\lambda_1, \ldots, \lambda_n$ as its entries on the main diagonal. Using the above observation, we find that we can premultiply this matrix by transvections to obtain another upper triangular matrix in $\mathrm{SL}(n, F)$ whose main diagonal entries are $\lambda_1, \ldots, \lambda_{n-2}, \lambda_{n-1}\lambda_n, 1$. By repeating this process, we obtain an upper triangular matrix in $\mathrm{SL}(n, F)$ whose main diagonal entries are $\lambda_1 \cdots \lambda_n, 1, \ldots, 1$; but this matrix has determinant 1, so we must have $\lambda_1 \cdots \lambda_n = 1$, and hence the matrix is upper unitriangular. This proves (2). ∎

PROPOSITION 3. The subgroups X_{ij} are conjugate in $\mathrm{SL}(n, F)$.

PROOF. Consider two root subgroups X_{ij} and $X_{i'j'}$. Let w be a permutation matrix sending $\mathbf{v_i}$ to $\mathbf{v_{i'}}$ and $\mathbf{v_j}$ to $\mathbf{v_{j'}}$; then we have $wX_{ij}(\alpha)w^{-1} = X_{i'j'}(\alpha)$ for any $\alpha \in F$ by part (v) of Lemma 4.6. Since the inverse of w is its transpose, we see that the determinant of w is ± 1. If $\det w = 1$, then $w \in \mathrm{SL}(n, F)$ and we are done. (Note that if $n > 3$, then by transposing columns if necessary we can always choose w such that $\det w = 1$.) Now suppose that $\det w = -1$, and let d be the matrix that differs from the identity matrix only by having -1 instead of 1 in the $(1, 1)$-entry. Then $dw \in \mathrm{SL}(n, F)$, and $dwX_{ij}(\alpha)(dw)^{-1} = dX_{i'j'}(\alpha)d^{-1} = X_{i'j'}(\pm\alpha)$ for any $\alpha \in F$, which completes the proof. ∎

Let Z be the subgroup of $\mathrm{GL}(n, F)$ consisting of the multiples of the identity matrix by elements of $F^{\times}$. The elements of Z are called the (non-zero) *scalar* matrices. We clearly have $Z \cong F^{\times}$.

PROPOSITION 4. The center of $\mathrm{GL}(n, F)$ is Z, and the center of $\mathrm{SL}(n, F)$ is $Z \cap \mathrm{SL}(n, F)$.

PROOF. If $n = 1$, then $\mathrm{GL}(1, F) \cong F^{\times} \cong Z$, and the result holds; hence we assume that $n > 1$. Let $G = \mathrm{GL}(n, F)$ or $\mathrm{SL}(n, F)$. We observe that $Z \cap G$ is contained in $Z(G)$; therefore, it suffices to show that any element of G that commutes with all elements of $\mathrm{SL}(n, F)$ lies in $Z \cap G$. Let $M = (m_{ij})$ be such a matrix. Let $1 \le i, j \le n$ be distinct, and consider $X_{ij}(1)$; by hypothesis we have $MX_{ij}(1) = X_{ij}(1)M$. By calculating and comparing the (i, i)- and (i, j)-entries of $MX_{ij}(1)$ and $X_{ij}(1)M$, we find that $m_{ji} = 0$ and $m_{ii} = m_{jj}$. We conclude that $m_{ij} = m_{11}\delta_{ij}$ for any i and j, and hence that $M \in Z \cap G$. (The same argument shows that the set of elements of $\mathcal{M}_n(F)$ which commute with all elements of $\mathcal{M}_n(F)$ is $\{\alpha I \mid \alpha \in F\} = Z \cup \{0\}$.) ∎

The group $\mathrm{GL}(n, F)/Z$ is called the *projective general linear group*, and the group $\mathrm{SL}(n, F)/Z \cap \mathrm{SL}(n, F)$ is called the *projective special linear group*. We denote these groups by $\mathrm{PGL}(n, F)$ and $\mathrm{PSL}(n, F)$, respectively. (Observe that the first isomorphism theorem implies that $\mathrm{PSL}(n, F) \cong Z\,\mathrm{SL}(n, F)/Z \le \mathrm{GL}(n, F)/Z = \mathrm{PGL}(n, F)$.) As $Z \cong F^{\times}$, we have $|\mathrm{PGL}(n, q)| = |\mathrm{SL}(n, q)|$ for any prime power q; however, to determine $|\mathrm{PSL}(n, q)|$ we must be able to count the number of nth roots of unity in the field of q elements.

PROPOSITION 5. Let q be a prime power. Then

$$|\text{PSL}(2,q)| = \begin{cases} q^3 - q & \text{if } 2 \mid q, \\ \frac{1}{2}(q^3 - q) & \text{if } 2 \nmid q. \end{cases}$$

PROOF. Any matrix in $Z \cap \text{SL}(2,q)$ must be of the form αI, where α is a square root of unity in the field of q elements. Such an element α is a root of the polynomial $X^2 - 1$, which has two distinct roots (namely 1 and -1) in any field whose characteristic is not equal to 2. Therefore, if $2 \nmid q$, then $|Z \cap \text{SL}(2,q)| = 2$. However, if $2 \mid q$, then $X^2 - 1 = (X - 1)^2$ has only one distinct root, and hence $|Z \cap \text{SL}(2,q)| = 1$. The result now follows via Proposition 1. ∎

Our main objective in this section is to show that $\text{PSL}(n, F)$ is simple whenever $n \geq 2$, except when $n = 2$ and $|F| = 2$ or 3. This result dates back to L. E. Dickson, another of the early architects of the mathematical tradition of the University of Chicago, who established it in his 1896 Ph.D. thesis for F a finite field; in 1893 E. H. Moore had established the result for F a finite field and $n = 2$, while in 1870 Camille Jordan had established the result for F a finite field of prime order and n arbitrary. We first need two lemmas.

LEMMA 6. If $n \geq 2$, then every transvection $X_{ij}(\alpha)$ is a commutator of elements of $\text{SL}(n, F)$, except when $n = 2$ and $|F| = 2$ or 3.

PROOF. If $n > 2$, then we have $X_{ij}(\alpha) = [X_{ik}(\alpha), X_{kj}(1)]$ by part (v) of Lemma 4.6, where k is unequal to either i or j. Now let $n = 2$. For any $\beta, \gamma \in F$ with $\beta \neq 0$, we observe that the commutator of $\begin{pmatrix} \beta & 0 \\ 0 & \beta^{-1} \end{pmatrix}$ and $\begin{pmatrix} 1 & \gamma \\ 0 & 1 \end{pmatrix}$ is $\begin{pmatrix} 1 & (\beta^2 - 1)\gamma \\ 0 & 1 \end{pmatrix}$. Therefore, $X_{12}(\alpha)$ will be expressible as a commutator in $\text{SL}(n, F)$ as long as there are $\beta \in F^{\times}$ and $\gamma \in F$ such that $\alpha = (\beta^2 - 1)\gamma$. If $|F| > 3$, then there is always some $\beta \in F^{\times}$ such that $\beta^2 \neq 1$, and we can then take $\gamma = \alpha(\beta^2 - 1)^{-1}$. The $X_{21}(\alpha)$ case is similar. ∎

Observe that Lemma 6 and Proposition 2 together imply that, except when $n = 2$ and $|F| = 2$ or 3, $\text{SL}(n, F)$ is its own derived group and is consequently also the derived group of $\text{GL}(n, F)$.

LEMMA 7. If $n \geq 2$, then the action of $\text{SL}(n, F)$ on the one-dimensional subspaces of $V_n(F)$ is doubly transitive.

PROOF. Let V_1, V_2, W_1, and W_2 be one-dimensional subspaces of $V_n(F)$, where $V_1 \neq V_2$ and $W_1 \neq W_2$, which are generated respectively by $\mathbf{c}_1, \mathbf{c}_2, \mathbf{d}_1$, and $\mathbf{d}_2$. Then $\mathbf{c}_1$ and $\mathbf{c}_2$ are linearly independent, as are $\mathbf{d}_1$ and $\mathbf{d}_2$, and consequently we can find $\mathbf{c}_3, \ldots, \mathbf{c}_n$ and $\mathbf{d}_3, \ldots, \mathbf{d}_n$ such that $\{\mathbf{c}_1, \ldots, \mathbf{c}_n\}$ and $\{\mathbf{d}_1, \ldots, \mathbf{d}_n\}$ are bases for $V_n(F)$. Let C (resp., D) be the $n \times n$ matrix whose ith column is $\mathbf{c}_i$ (resp., $\mathbf{d}_i$) for each $1 \leq i \leq n$; clearly, C and D are matrices of rank n and hence lie in $\mathrm{GL}(n, F)$. Let $e = \det D / \det C$, and let E be the matrix differing from the identity matrix only by having e instead of 1 in the $(1,1)$-entry. Then $\det DE^{-1}C^{-1} = 1$, and $DE^{-1}C^{-1}$ sends $\mathbf{c}_1$ to $e^{-1}\mathbf{d}_1$ and $\mathbf{c}_2$ to $\mathbf{d}_2$; hence we have found an element of $\mathrm{SL}(n, F)$ sending V_1 to W_1 and V_2 to W_2, as required. ∎

THEOREM 8. If $n \geq 2$, then $\mathrm{PSL}(n, F)$ is simple, except when $n = 2$ and $|F| = 2$ or 3.

PROOF. Let $S = \mathrm{SL}(n, F)$, and let $P \leqslant S$ be the stabilizer of $F\mathbf{v}_1$ under the action of S on the one-dimensional subspaces of $V_n(F)$. By Lemma 7 and Proposition 3.8, P is a maximal subgroup of S. Let K be the set of upper unitriangular matrices whose only non-zero entries outside of the main diagonal occur in the first row. We find that K is an abelian normal subgroup of P.

Let $N \lhd S$. By the correspondence theorem and the definition of $\mathrm{PSL}(n, F)$, it suffices to show that N is comprised of scalar matrices. Suppose first that $N \leqslant P$. Then N stabilizes $F\mathbf{v}_1$, and hence $N = sNs^{-1}$ stabilizes $s(F\mathbf{v}_1)$ for any $s \in S$ by Lemma 3.2. Since the action of S on the one-dimensional subspaces of $V_n(F)$ is transitive by Lemma 7, this shows that N stabilizes every one-dimensional subspace of $V_n(F)$. In particular, N stabilizes each $F\mathbf{v}_i$, which shows that the elements of N are diagonal matrices. But N also stabilizes each $F(\mathbf{v}_i + \mathbf{v}_j)$, which shows that the elements of N must be scalar matrices.

Now suppose that $N \not\leqslant P$. Then we have $P < PN \leqslant S$, which forces $PN = S$ since P is maximal in S. Let $\eta: S \to S/N$ be the natural map. We have $\eta(P) = PN/N = S/N = \eta(S)$ and $\eta(K) = KN/N = \eta(KN)$. As $K \unlhd P$, it follows that $\eta(K) \unlhd \eta(P)$, and hence that $\eta(KN) \unlhd \eta(S)$; the correspondence theorem now gives $KN \unlhd S$. Observing that K is the group generated by the root subgroups $X_{12}, \ldots, X_{1n}$, we see that all conjugates in S of these root subgroups lie in KN. But we see via Proposition 3 that these conjugates include all of the X_{ij}; as the X_{ij} generate S by Proposition 2, we have $KN = S$. By the first isomorphism theorem, we now have $S/N = KN/N \cong K/K \cap N$, and since K is abelian it follows that S/N is also abelian. Therefore, all commutators of elements of S lie in N by Proposition 2.6, and in particular all matrices $X_{ij}(\alpha)$ lie in N by Lemma 6. Thus N contains every X_{ij}, and Proposition 2 now implies that $N = S$, which is a contradiction. ∎

EXERCISES

1. Let B be the standard Borel subgroup of $\mathrm{GL}(n, F)$. Determine all subgroups of $\mathrm{SL}(n, F)$ which contain $B \cap \mathrm{SL}(n, F)$.
2. Show that $\mathrm{PSL}(2,2) \cong \Sigma_3$ and $\mathrm{PSL}(2,3) \cong A_4$, and that these groups are not simple.
3. Show that $\mathrm{PSL}(2,4)$ and $\mathrm{PSL}(2,5)$ are both isomorphic with A_5.
4. Show that $\mathrm{PSL}(2,7) \cong \mathrm{GL}(3,2)$.
5. Show that $\mathrm{PSL}(2,9) \cong A_6$.
6. By Exercise 3 and Theorem 8, the group A_5 is simple. Attempt to prove this fact directly by mimicking the proof of Theorem 8, with A_5 in place of S, A_4 in place of P, and the Klein four-group in place of K.

FURTHER EXERCISES

Let F be a field, and let $V = F^2$. We define an equivalence relation on $V - \{0\}$ by $\mathbf{v} \sim \mathbf{w}$ iff $\mathbf{v} = \alpha \mathbf{w}$ for some $\alpha \in F^\times$. The set of equivalence classes under this relation is called the *one-dimensional projective space* (or *projective line*) over F and is denoted by $\mathbb{P}^1(F)$. We write elements of V as column vectors, and the element of $\mathbb{P}^1(F)$ that is the equivalence class of $\binom{x}{y}$ shall be written as $\left[\begin{smallmatrix} x \\ y \end{smallmatrix}\right]$.

7. Show that there is a well-defined action of $\mathrm{PSL}(2, F)$ on $\mathbb{P}^1(F)$ given by
$$\overline{\begin{pmatrix} a & b \\ c & d \end{pmatrix}} \begin{bmatrix} x \\ y \end{bmatrix} = \begin{bmatrix} ax + by \\ cx + dy \end{bmatrix}$$
and that this action is doubly transitive. (Here $\overline{\left(\begin{smallmatrix} a & b \\ c & d \end{smallmatrix}\right)}$ denotes the image in $\mathrm{PSL}(2, F)$ of $\left(\begin{smallmatrix} a & b \\ c & d \end{smallmatrix}\right) \in \mathrm{SL}(2, F)$.)
8. (cont.) Determine the proper definition of n-dimensional projective space $\mathbb{P}^n(F)$ for arbitrary n, and show that $\mathrm{PSL}(n + 1, F)$ acts doubly transitively on $\mathbb{P}^n(F)$.
9. Let F be the field of 7 elements. For $0 \leq i \leq 6$, let i represent $\left[\begin{smallmatrix} 1 \\ i \end{smallmatrix}\right] \in \mathbb{P}^1(F)$; denote the remaining element $\left[\begin{smallmatrix} 0 \\ 1 \end{smallmatrix}\right]$ of $\mathbb{P}^1(F)$ by ∞.

View Σ_8 as being the group of permutations of $\{\infty, 0, 1, 2, 3, 4, 5, 6\}$, and let $\varphi \colon \mathrm{PSL}(2,7) \to \Sigma_8$ be the monomorphism arising (as in Proposition 3.1) from the action of $\mathrm{PSL}(2,7)$ on $\mathbb{P}^1(F)$. Show that the image of φ is the subgroup of A_8 generated by (0 1 2 3 4 5 6), (1 4 2)(3 5 6), and $(\infty$ 0)(1 3)(2 5)(4 6). (HINT: Show first that $\mathrm{PSL}(2,7)$ is generated by the images under the natural map of the elements $\left(\begin{smallmatrix} 1 & 0 \\ 1 & 1 \end{smallmatrix}\right)$, $\left(\begin{smallmatrix} 4 & 0 \\ 0 & 2 \end{smallmatrix}\right)$, and $\left(\begin{smallmatrix} 0 & 3 \\ 2 & 0 \end{smallmatrix}\right)$ of $\mathrm{SL}(2,7)$.) For the relevance of this exercise, see Exercises 7.13–17.

Recall the definition of a BN-pair from the further exercises to Section 4.

10. Let $G = \mathrm{GL}(n, F)$. Let B be the standard Borel subgroup of G, let N be the subgroup of G of monomial matrices, and let $T = B \cap N$ be the subgroup of diagonal matrices. Let $B_0 = B \cap \mathrm{SL}(n, F)$, let $N_0 = N \cap \mathrm{SL}(n, F)$, and let $T_0 = T \cap \mathrm{SL}(n, F) = B_0 \cap N_0$. Show that B_0 and N_0 form a BN-pair of $\mathrm{SL}(n, F)$, with the associated Weyl group $W_0 = N_0/T_0$ being isomorphic with Σ_n.

11. Let $P = \mathrm{PSL}(n, F)$, and let $\overline{B_0}, \overline{N_0}$, and $\overline{T_0}$ be the images in P of the subgroups B_0, N_0, and T_0 of $\mathrm{SL}(n, F)$. Show that $\overline{B_0}$ and $\overline{N_0}$ form a BN-pair of P (with again the Weyl group $\overline{W_0} = \overline{N_0}/\overline{T_0}$ being isomorphic with Σ_n).

3
Local Structure

In many branches of mathematics, it is profitable to study an issue by somehow "localizing" with respect to a given prime number. In this chapter, we adapt this doctrine to group theory by studying finite groups through their subgroups of prime-power order. This notion of looking at the "local structure" of finite groups has proven to be very powerful. We start in Section 7 with Sylow's theorem on subgroups of maximal prime-power order. Section 8 concentrates on the properties of finite groups of prime-power order. Section 9 gives an important application, the Schur-Zassenhaus theorem.

7. Sylow's Theorem

Throughout this section, we let G denote a finite group and p a prime divisor of $|G|$. We use $|G|_p$ to denote the highest power of p that divides $|G|$, so that $|G|_p = p^n$ where $n \in \mathbb{N}$ is such that p^n divides $|G|$ but p^{n+1} does not.

We say that $g \in G$ is a *p-element* if its order is a power of p. We say that G is a *p-group* if $|G|$ is a power of p, and that $H \leqslant G$ is a *p-subgroup* of G if $|H|$ is a power of p. Every element of a p-group is a p-element. (Indeed, an infinite group is said to be a p-group iff every element is a p-element.) We say that $H \leqslant G$ is a *Sylow p-subgroup* of G if H is a p-subgroup of order $|G|_p$, which is of course the maximal possible order of a p-subgroup of G. A subgroup of G is said to be a *Sylow subgroup* of G if it is a Sylow p-subgroup for some prime divisor p of G.

For example, let $n \in \mathbb{N}$ and let $G = \mathrm{GL}(n, p)$. We see from Proposition 4.2 that $|G| = p^{\frac{n(n-1)}{2}}(p^n - 1) \cdots (p - 1)$, and hence that $|G|_p = p^{\frac{n(n-1)}{2}}$. Let U be the subgroup of G consisting of all upper uni-triangular matrices. It is straightforward to show that $|U| = p^{n-1}p^{n-2} \cdots p^2 p = p^{\frac{n(n-1)}{2}} = |G|_p$; therefore U is a Sylow p-subgroup of G, as is any conjugate of U. Now let x be a non-trivial p-element of G, so that $x^{p^a} - I = 0$ for some $a \in \mathbb{N}$. Since the entries of x lie in the field $\mathbb{Z}/p\mathbb{Z}$ of characteristic p, we may rewrite this equation as $(x - I)^{p^a} = 0$. Thus, the minimal polynomial of x divides $(X - 1)^{p^a}$, and since the minimal and characteristic polynomials have the same irreducible factors, the characteristic polynomial of x must be $(X - 1)^n$. Therefore, all p-elements of G are unipotent, and hence any p-subgroup of G is unipotent. By Kolchin's theorem, we conclude that any p-subgroup of G is conjugate with a subgroup of the Sylow p-subgroup U. In particular, any Sylow p-subgroup of G is conjugate with U.

We just established that $\mathrm{GL}(n, p)$ has a Sylow p-subgroup, that all the Sylow p-subgroups of $\mathrm{GL}(n, p)$ are conjugate, and that any p-subgroup of $\mathrm{GL}(n, p)$ is contained in a Sylow p-subgroup. The following result, which dates back to 1871 and is fundamental to the study of finite groups, shows that these properties hold in any finite group G for any prime divisor p of $|G|$.

SYLOW'S THEOREM.   (i) G has at least one Sylow p-subgroup.
  (ii) All the Sylow p-subgroups of G are conjugate.
(iii) Any p-subgroup of G is contained in a Sylow p-subgroup.
(iv) The number of Sylow p-subgroups of G is congruent to 1 modulo p.

PROOF. Let $|G| = p^n m$, where $p^n = |G|_p$ and hence $p \nmid m$. Let X be the collection of all subsets of G having $|G|_p$ elements; X is a G-set under left multiplication.

Suppose that there is an orbit $\mathcal{O}$ of X such that $p \nmid |\mathcal{O}|$. Let $A \in \mathcal{O}$ be such that $1 \in A$, and let $P \leqslant G$ be the stabilizer of A. Since $1 \in A$, we have $P \subseteq PA = A$, and thus $|P| \leq |A| = |G|_p$. We also have $|\mathcal{O}| = |G : P|$ by Corollary 3.5, and hence $|G| = |P||\mathcal{O}|$; since $p \nmid |\mathcal{O}|$, this shows that $|G|_p$ divides $|P|$. Therefore $|P| = |G|_p$, and so P is a

Sylow p-subgroup of G. It follows that $A = P$ and that $\mathcal{O}$ is the coset space G/P. Conversely, suppose that P is a Sylow p-subgroup of G. The coset space G/P is a collection of subsets of G of order $|G|_p$ and hence is contained in X; moreover, G/P is the orbit in X of P, and p does not divide $|G/P| = m$. Thus, we have a bijective correspondence between the set of Sylow p-subgroups of G and the set of orbits of X whose cardinality is coprime to p, where any such orbit is the coset space of the corresponding Sylow p-subgroup.

Let X' be the set of elements of X that lie in an orbit whose cardinality is not divisible by p. Since p must divide $|X - X'|$, we have $|X| \equiv |X'|$ (mod p). If $S \in X'$, then the orbit of X containing S is, as noted above, the set of cosets of a Sylow p-subgroup of G and hence has cardinality m. If we now let r be the number of Sylow p-subgroups of G, then since r also equals the number of orbits of X contained in X', we have $rm = |X'| \equiv |X| = \binom{p^n m}{p^n}$ (mod p). Since $p \nmid m$, this implies that the value of r, modulo p, depends only on the order of G and not on G itself; that is, any two groups of the same order have the same number, modulo p, of Sylow p-subgroups. But we see from Theorem 1.4 that the cyclic group of order $p^n m$ has exactly 1 subgroup of order $|G|_p$. Therefore $r \equiv 1$ (mod p), proving (iv), and in particular $r > 0$, proving (i).

Now let P be some Sylow p-subgroup of G, and let Q be an arbitrary p-subgroup of G. The group Q acts on the set Y of conjugates of P in G by conjugation; the action of $x \in Q$ sends $gPg^{-1} \in Y$ to $x(gPg^{-1})x^{-1} = (xg)P(xg)^{-1}$. The cardinality of each orbit, being the index of a subgroup of the p-group Q, is some (perhaps trivial) power of p. We have $|Y| = |G : N_G(P)|$ by Proposition 3.14; using factorization of indices, we see that $|Y|$ divides $|G : P| = m$, and since $p \nmid m$ we must have $p \nmid |Y|$. Thus, there must be some orbit of Y containing only one element, as otherwise p would divide $|Y|$.

Let $\{P_1\}$ be a single-element orbit of Y under the action of Q. Then we have $xP_1x^{-1} = P_1$ for all $x \in Q$; consequently $QP_1 = P_1Q$, and so $QP_1 \leqslant G$ by Proposition 1.3. Clearly $|P_1| \leq |QP_1|$; but since $|QP_1| = |P_1||Q : Q \cap P_1|$ by Proposition 3.12, we see that QP_1 must be a p-group. Therefore $|QP_1| = |P_1|$, which forces $Q = Q \cap P_1$ and hence $Q \leqslant P_1$. Since P_1 is a Sylow p-subgroup of G, this proves (iii). If we now take Q to be a Sylow p-subgroup of G, then as $Q \leqslant P_1$ and $|Q| = |P_1|$, we must have $Q = P_1$; in particular Q is conjugate with P, which proves (ii). ∎

COROLLARY 1. The number of Sylow p-subgroups of G divides $|G|/|G|_p$.

PROOF. Let r be the number of Sylow p-subgroups of G, and let P be some such subgroup. By part (ii) of Sylow's theorem, r equals the number of conjugates of P in G; since this number equals $|G : N_G(P)|$ by Proposition 3.14, we see via factorization of indices that r divides $|G : P| = |G|/|G|_p$. ■

Sylow's theorem immediately implies the following classical result.

CAUCHY'S THEOREM. G has an element of order p.

PROOF. By part (i) of Sylow's theorem, G has a non-trivial Sylow p-subgroup, and hence G has a non-trivial p-element, some power of which is of order p. ■

(Other proofs of Sylow's theorem often assume that Cauchy's theorem is already known, as was the case historically.)

The next result gives a relationship between the Sylow p-subgroups of a group and those of its normal subgroups and quotient groups.

PROPOSITION 2. Let $N \trianglelefteq G$, and let P be a Sylow p-subgroup of G. Then PN/N is a Sylow p-subgroup of G/N, and $P \cap N$ is a Sylow p-subgroup of N.

PROOF. We have $|G/N : PN/N| = |G : PN|$ by the correspondence theorem. But since $|G : PN|$ divides $|G : P|$ and $p \nmid |G : P|$, we see that $p \nmid |G/N : PN/N|$. Since $PN/N \cong P/P \cap N$ by the first isomorphism theorem, PN/N must be a p-group; it now follows that PN/N is a Sylow p-subgroup of G/N.

It follows from Proposition 3.12 that $|N : P \cap N| = |PN : P|$; but since PN is a subgroup of G by Proposition 1.7, and P is a Sylow p-subgroup of G, we must have $p \nmid |PN : P|$. Therefore, $P \cap N$ is a p-subgroup of N whose index in N is coprime to p, as required. ■

If $H \leqslant G$ and P is a Sylow p-subgroup of G, then $P \cap H$ need not be a Sylow p-subgroup of H; the above proof falls apart here since PH may not be a subgroup of G. However, if Q is a Sylow p-subgroup of H, then by part (iii) of Sylow's theorem Q is contained in some Sylow p-subgroup P' of G, and we must have $Q = P' \cap H$. Part (ii) of Sylow's theorem now implies that P' and P are conjugate in G. Therefore, there is some $g \in G$ such that $gPg^{-1} \cap H$ is a Sylow p-subgroup of H.

In the remainder of this section, we present some typical applications of Sylow's theorem to the study of finite groups.

PROPOSITION 3. Let p and q be distinct primes, with $p > q$. If $p \not\equiv 1 \pmod{q}$, then any group of order pq is isomorphic with $\mathbf{Z_{pq}}$. If $p \equiv 1 \pmod{q}$, then any abelian group of order pq is isomorphic with $\mathbf{Z_{pq}}$, and there is exactly one isomorphism class of non-abelian groups of order pq.

PROOF. Let G be a group of order pq, let P be a Sylow p-subgroup of G, and let Q be a Sylow q-subgroup of G. Since $|P| = p$ and $|Q| = q$, we have $P \cong \mathbf{Z_p}$ and $Q \cong \mathbf{Z_q}$. Lagrange's theorem gives $P \cap Q = 1$, and hence it follows from Proposition 3.12 that $G = PQ$.

By Sylow's theorem and Corollary 1, the number of conjugates of P in G divides $|G : P| = q$ and is congruent to 1 modulo p. But we have $q \not\equiv 1 \pmod{p}$ since $p > q$; therefore, P must have only 1 conjugate in G, and hence $P \trianglelefteq G$. We can similarly show that Q has either 1 or p conjugates in G, and that the latter case can only occur when $p \equiv 1 \pmod{q}$. If Q has only 1 conjugate in G, then $Q \trianglelefteq G$, and consequently $G = P \times Q \cong \mathbf{Z_p} \times \mathbf{Z_q} \cong \mathbf{Z_{pq}}$ via Lemma 2.8. This will be the case if $p \not\equiv 1 \pmod{q}$, or if $p \equiv 1 \pmod{q}$ and G is abelian.

Now suppose that Q has p conjugates in G, in which case G is non-abelian and $p \equiv 1 \pmod{q}$. We have $P \trianglelefteq G$, $G = PQ$, and $P \cap Q = 1$, which gives $G = P \rtimes Q$. Let $\varphi : Q \to \mathrm{Aut}(P)$ be the conjugation homomorphism. If $\ker \varphi \neq 1$, then as Q is simple we must have $\ker \varphi = Q$, in which case φ is trivial and hence G is abelian. Therefore φ must be injective.

We conclude that if $p \equiv 1 \pmod{q}$ and there is a non-abelian group of order pq, then there is a monomorphism from $\mathbf{Z_q}$ to $\mathrm{Aut}(\mathbf{Z_p})$. Conversely, given a monomorphism $\varphi : \mathbf{Z_q} \to \mathrm{Aut}(\mathbf{Z_p})$, we can construct a non-abelian group of order pq, namely $\mathbf{Z_p} \rtimes_\varphi \mathbf{Z_q}$. To complete the proof, we must exhibit a monomorphism $\varphi : \mathbf{Z_q} \to \mathrm{Aut}(\mathbf{Z_p})$, which shows the existence of a non-abelian group of order pq; we must then show that if $\psi : \mathbf{Z_q} \to \mathrm{Aut}(\mathbf{Z_p})$ is another such monomorphism, then the groups $\mathbf{Z_p} \rtimes_\varphi \mathbf{Z_q}$ and $\mathbf{Z_p} \rtimes_\psi \mathbf{Z_q}$ are isomorphic.

We see from Proposition 2.2 that $\mathrm{Aut}(\mathbf{Z_p}) \cong \mathbf{Z_{p-1}}$. Now $q \mid (p-1)$ by hypothesis, so from Theorem 1.4 we see that $\mathrm{Aut}(\mathbf{Z_p})$ has a unique subgroup K of order q. Using the characterization of $\mathrm{Aut}(\mathbf{Z_p})$ given in Proposition 2.1, we see that there is some $1 < r < p$ such that K is generated by the automorphism σ_r sending every element to its rth power. We define $\varphi\colon \mathbf{Z_q} \to \mathrm{Aut}(\mathbf{Z_p})$ by letting φ send a generator of $\mathbf{Z_q}$ to σ_r. Then φ is a monomorphism with image K, which proves existence. Now if $\psi\colon \mathbf{Z_q} \to \mathrm{Aut}(\mathbf{Z_p})$ is another monomorphism, then by the uniqueness of K we must have $\psi(\mathbf{Z_q}) = K = \varphi(\mathbf{Z_q})$; we now have $\mathbf{Z_p} \rtimes_\varphi \mathbf{Z_q} \cong \mathbf{Z_p} \rtimes_\psi \mathbf{Z_q}$ by Proposition 2.11, as required. ∎

We will now demonstrate the use of Sylow's theorem as a tool in the study of finite simple groups. One consequence of part (iii) of Sylow's theorem which will be of particular use is that every p-element of G is contained in a Sylow p-subgroup of G; this will allow us to use information about the Sylow p-subgroups of G to count the number of p-elements of G.

THEOREM 4. A_5 is simple.

PROOF. We have $|A_5| = 5!/2 = 60 = 2^2 \cdot 3 \cdot 5$. We see from Sylow's theorem and Corollary 1 that the number of Sylow 5-subgroups of A_5 divides $60/5 = 12$ and is congruent to 1 modulo 5; using the fact that any 5-cycle generates a Sylow 5-subgroup, we observe that this number is not 1. Thus A_5 has 6 Sylow 5-subgroups. As no two of these subgroups can have an element of order 5 in common, we conclude that A_5 has $6 \cdot (5-1) = 24$ elements of order 5. Similarly, A_5 could have 1, 4, or 10 Sylow 3-subgroups, and by inspection the number of Sylow 3-subgroups exceeds 4; consequently, A_5 has $10 \cdot (3-1) = 20$ elements of order 3.

Let $1 \le i \le 5$, let $\{a,b,c,d\}$ be the complement in $\{1,2,3,4,5\}$ of $\{i\}$, and let $V_i = \{1, (a\ b)(c\ d), (a\ c)(b\ d), (a\ d)(b\ c)\}$. We see easily that each V_i is a Sylow 2-subgroup of A_5 and that if $i \ne j$ then $V_i \cap V_j = 1$. By inspection, we see that $\rho V_i \rho^{-1} = V_{\rho(i)}$ for any $\rho \in A_5$. It now follows from part (ii) of Sylow's theorem that $V_1, \dots, V_5$ are the only Sylow 2-subgroups of A_5. It also follows that A_5 has $5 \cdot (4-1) = 15$ elements of order 2 and that every element of order 2 in A_5 is conjugate with some other element of order 2.

Now suppose that N is a proper normal subgroup of A_5, and let $n = |N| \le 30$. Suppose that $5 \mid n$. Then N contains a Sylow 5-subgroup of A_5; but as N is normal in G, N also contains all conjugates of that Sylow 5-subgroup and hence all 6 Sylow 5-subgroups of A_5. In particular, N contains 24 elements of order 5, which forces $n = 30$. Now $3 \mid 30$, so N contains 1, and hence all 10, Sylow 3-subgroups of A_5; thus N contains 20 elements of order 3, which is a contradiction. Therefore $5 \nmid n$, which gives $n \leq 12$. If $3 \mid n$, then by the argument just given, N contains 20 elements of order 3, which is a contradiction; therefore $n = 1$, 2, or 4. If $n = 4$, then N is a Sylow 2-subgroup of A_5 and hence has 5 conjugates in A_5, which contradicts the normality of N. But we observed previously that each element of order 2 in A_5 has some conjugate other than itself; consequently A_5 cannot have a normal subgroup of order 2. Therefore $n = 1$, which proves that A_5 is simple. ∎

THEOREM 5. Any simple group of order 60 is isomorphic with A_5.

PROOF. Let G be a simple group of order 60. As observed after Proposition 3.1, if G has a subgroup of index n, then there is a monomorphism from G to Σ_n associated with the action of G via left multiplication on the coset space of that subgroup. Since $|G| > |\Sigma_n|$ when $n < 5$, we conclude that G cannot have a proper subgroup of index less than 5. We shall now show that G does have a subgroup of index exactly 5 and consequently that G is isomorphic with a subgroup of Σ_5.

Suppose that G does not have a subgroup of index 5. By Corollary 1, the number of Sylow 2-subgroups of G divides $60/4 = 15$; since this number is equal to the index in G of the normalizer of a Sylow 2-subgroup, by our hypothesis and the previous paragraph it must be 15. Let S_1 and S_2 be distinct Sylow 2-subgroups of G, and suppose that there is some $1 \neq t \in S_1 \cap S_2$. Then $|C_G(t)| > 4$ since S_1 and S_2 are distinct and abelian, and 4 divides $|C_G(t)|$ since $S_1 \leq C_G(t)$; hence we must have $|G : C_G(t)| \leq 5$. Our hypothesis and the previous paragraph force $C_G(t) = G$; but this gives $t \in Z(G)$, which contradicts the simplicity of G. Therefore, no two of the 15 Sylow 2-subgroups of G have a non-trivial element in common, and hence G has $15 \cdot (4 - 1) = 45$ elements of order 2 or 4. But G is simple, so G must have more than one Sylow 5-subgroup; thus G has 6 Sylow 5-subgroups and therefore has 24 elements of order 5. This gives a contradiction. Consequently, we can conclude that G does have a subgroup of index 5, and hence that G is isomorphic with a subgroup of Σ_5.

We now identify G with its isomorphic copy inside Σ_5. Since $|\Sigma_5 : G| = 2$, we have $G \trianglelefteq \Sigma_5$ by Proposition 1.8. Suppose that $G \neq A_5$. Then $|GA_5| > 60$, which forces $GA_5 = \Sigma_5$. Proposition 3.10

now gives

$$|G \cap A_5| = \frac{|G||A_5|}{|GA_5|} = \frac{60 \cdot 60}{120} = 30,$$

and hence $|G : G \cap A_5| = 2$. We now have $1 \neq G \cap A_5 \lhd G$ by Proposition 1.8, which contradicts the simplicity of G. Therefore $G = A_5$. (Observe that this argument can also be used to show that A_5 is the only non-trivial proper normal subgroup of Σ_5.) ∎

COROLLARY 6. $\mathrm{PSL}(2,4) \cong \mathrm{PSL}(2,5) \cong A_5$.

PROOF. We have $|\mathrm{PSL}(2,4)| = |\mathrm{PSL}(2,5)| = 60$ by Proposition 6.5; but $\mathrm{PSL}(2,4)$ and $\mathrm{PSL}(2,5)$ are simple by Theorem 6.8, and so the result follows from Theorem 5. ∎

It is a relatively easy exercise to show that the only simple groups of order less than 60 are the cyclic groups of prime order, and hence that A_5 is the non-abelian finite simple group having smallest order. The non-abelian finite simple group of next smallest order is $\mathrm{PSL}(2,7)$, which has order 168.

EXERCISES

Throughout these exercises, G is a finite group and p is a prime divisor of $|G|$.

1. Let $H \leqslant G$ and let P be a Sylow p-subgroup of G. Prove, without reference to Sylow's theorem, that there is some conjugate of P whose intersection with H is a Sylow p-subgroup of H.

2. (cont.) Use Exercise 1 to give an alternate proof of part (i) of Sylow's theorem.

3. Give an alternate proof of parts (ii) and (iii) of Sylow's theorem by considering the action of an arbitrary p-subgroup Q of G on the coset space G/P, where P is a Sylow p-subgroup of G.

4. Prove the following generalization of part (iv) of Sylow's theorem: If $|G|$ is divisible by p^b, and $H \leqslant G$ has order p^a where $a \leq b$, then the number of subgroups of G that both contain H and have order p^b is congruent to 1 modulo p.

5. Show that if P is a Sylow p-subgroup of G, then $N_G(P)$ is a self-normalizing subgroup of G, meaning that $N_G(N_G(P)) = N_G(P)$. (Any subgroup of G that can be written as $N_G(P)$ for some non-trivial p-subgroup P of G is said to be a p-local subgroup of G.)

6. Given a prime p, find an example of a finite group having exactly $1 + p$ Sylow p-subgroups. Can this be done for $1 + 2p$? $1 + 3p$?

7. Show that if G has exactly $1 + kp$ Sylow p-subgroups for some $k \in \mathbb{N}$, then there is a subgroup of Σ_{1+kp} having exactly $1 + kp$ Sylow p-subgroups.

8. Let $n \geq 6$, and assume by induction that A_{n-1} is simple. Let N be a non-trivial proper normal subgroup of A_n.

(a) Show that $A_n = N \rtimes A_{n-1}$.

(b) Let $N^{\#}$ be the set of non-identity elements of N, and consider the action of A_{n-1} on $N^{\#}$ given by the conjugation homomorphism from A_{n-1} to $\mathrm{Aut}(N)$. Show that $N^{\#}$ is isomorphic as an A_{n-1}-set to $\{1, \ldots, n-1\}$.

(c) Show that A_{n-1} acts triply transitively on $N^{\#}$. (An action of a group G on a set X is *triply transitive* if for every pair of triples (x_1, x_2, x_3) and (y_1, y_2, y_3), where $x_i, y_i \in X$ for all i and $x_i \neq x_j$ (resp., $y_i \neq y_j$) when $i \neq j$, there is some $g \in G$ such that $gx_i = y_i$ for all i.)

(d) Derive a contradiction, and conclude that A_n is simple.

The remaining exercises develop a proof that if G is a finite simple group with $|G| \leq 200$, then either $G \cong \mathbf{Z_p}$ for some prime p, or $G \cong A_5$, or $G \cong \mathrm{PSL}(2,7)$. This was established by Hölder in 1892. We will need to assume two facts, which will be proved in later sections:

A. If $|G| = p^n$ where $n > 1$, then G is not simple (Section 8).

B. If G is a p-group and $H < G$, then $H < N_G(H)$ (Section 11).

9. Show that G is not simple whenever one of the following statements is true, where p is an odd prime:

(a) $|G| = 2^m p^n$, where $2^k \not\equiv 1 \pmod{p}$ for any $1 \leq k \leq m$.

(b) $|G| = p^n q$, where $q \neq p$ is prime and $q \not\equiv 1 \pmod{p}$.

(c) $|G| = ap$, where $p \nmid a$ and $kp + 1 \nmid a$ for any $k \in \mathbb{N}$.

10. Show that G is not simple whenever one of the following statments is true:

(a) $|G| = p^a(p+1)$, where $a > 1$.

(b) $|G| = p^a(p+3)$, where $a > 3$ if $p = 2$ and $a > 1$ if $p > 3$.

(c) $|G| = p^a(p^2 - 1)$, where $a > 1$ and p is odd.

11. Show that if $|G| \in \{30, 56, 105, 132\}$, then G is not simple.

12. Show that G is not simple in the following cases:

(a) $|G| = 90$. (HINT: Show that if Q is a Sylow 3-subgroup of G and $1 \neq x \in Q$, then $C_G(x) = Q$.)

(b) $|G| = 112$. (HINT: Use fact B above to show that any two Sylow 2-subgroups of G intersect trivially.)

(c) $|G| = 120$. (HINT: Show that G imbeds in Σ_6, and contradict Exercise 3.8.)

(d) $|G| = 144$. (HINT: Argue as in part (i) above.)

(e) $|G| = 180$. (HINT: Show that if Q is a Sylow 3-subgroup of G and $1 \neq x \in Q$, then $|C_G(x) : Q| \leq 2$.)

From fact A above and Exercises 9–12, we can conclude that if G is a simple group with $|G| \leq 200$, then $|G|$ either is prime or is equal to either 60 or 168. (Verify this.) In light of Theorem 5 and Theorem 6.8, all we need now show is that any two simple groups of order $168 = 2^3 \cdot 3 \cdot 7$ are isomorphic. We accomplish this in Exercises 13–17 below by showing that any simple group of order 168 is isomorphic with a specific subgroup of A_8. (Compare with Exercise 6.9.) Let G be a simple group of order 168, let $P = <x>$ be a Sylow 7-subgroup of G, and let $H = N_G(P)$.

13. Show that $H = \{g \in G \mid gxg^{-1} \in P\}$ is a non-abelian group of order 21 which is generated by x and y, where y has order 3 and $yxy^{-1} = x^4$.

14. (cont.) Show that $N_G(<y>)$ is isomorphic with Σ_3 and is generated by y and z, where z has order 2 and $yzy = z$.

15. (cont.) Show that $G/H = \{H, zH, xzH, \ldots, x^6zH\}$. Conclude that $G = <x, y, z>$.

Let $\varphi: G \to \Sigma_8$ be the monomorphism corresponding to the action of G on the coset space G/H. Observe that $\varphi(G) = \varphi(G)' \leqslant \Sigma_8' \leqslant A_8$. View Σ_8 as the group of permutations of the set $\{\infty, 0, 1, 2, 3, 4, 5, 6\}$; for each $0 \leq i \leq 6$, associate i with the coset $x^i z H$, and associate ∞ with the coset H.

16. (cont.) Show that $\varphi(x) = (0\ 1\ 2\ 3\ 4\ 5\ 6)$, that $\varphi(y) = (1\ 4\ 2)(3\ 5\ 6)$, and that $\varphi(z) = z_i$ for some i, where $z_1 = (\infty\ 0)(1\ 3)(2\ 5)(4\ 6)$, $z_2 = (\infty\ 0)(1\ 5)(2\ 6)(3\ 4)$, and $z_3 = (\infty\ 0)(1\ 6)(2\ 3)(4\ 5)$.

17. (cont.) Let $L_i = <\varphi(x), \varphi(y), z_i> \leqslant A_8$ for $i = 1, 2, 3$. Show that $L_1 = L_2 = L_3$. Conclude that G is uniquely determined up to isomorphism.

8. Finite p-groups

We have seen that Sylow's theorem allows us to gain information about finite groups by concentrating on the properties of their subgroups of prime-power order. This motivates the study of finite p-groups, a rich subject whose surface we shall but scratch in this section. Our first result is the most basic property of finite p-groups.

THEOREM 1. If P is a non-trivial finite p-group, then $Z(P)$ is also non-trivial. Moreover, if $1 \neq N \trianglelefteq P$, then $N \cap Z(P) \neq 1$.

PROOF. Suppose that $Z(P) = 1$. Let $K_1 = \{1\}, K_2, \ldots, K_r$ be the conjugacy classes of P. If $K_i = \{x\}$ for some i and some $x \in P$, then $gxg^{-1} = x$ for all $g \in P$, which shows that $x \in Z(P) = 1$. Therefore, $|K_i| > 1$ for each $1 < i \leq r$. But each K_i is an orbit under the action of P on itself by conjugation (as in Proposition 3.13), and thus $|K_i|$ divides $|P|$ for all i by Corollary 3.5; since P is a p-group, we must have $|K_i| \equiv 0 \pmod{p}$ for each $1 < i \leq r$. We now have $|P| = 1 + |K_2| + \cdots + |K_r| \equiv 1 \pmod{p}$, since P is the disjoint union of $K_1, \ldots, K_r$. This is a contradiction, since $|P|$ is divisible by p. Therefore, $Z(P)$ is non-trivial. Now let $1 \neq N \trianglelefteq P$. Then N is the disjoint union of the K_j for all j in some subset S of $\{1, \ldots, r\}$, where S contains 1 and $|S| > 1$. If $N \cap Z(P) = 1$, then we would have $|K_j| > 1$ for all $1 \neq j \in S$, which as above would lead to the contradiction that $|N| \equiv 1 \pmod{p}$. ∎

In particular, this implies that the only finite simple p-groups are the groups of prime order, as was mentioned in the exercises to Section 7.

The center of a non-abelian finite p-group cannot be trivial, but neither can it be too large:

LEMMA 2. If G is a non-abelian group, then $G/Z(G)$ cannot be cyclic. In particular, if P is a finite p-group, then $|P : Z(P)| \neq p$.

PROOF. Suppose that $G/Z(G) = <xZ(G)>$ for some $x \in G$. Then G would be generated by the set $S = Z(G) \cup \{x\}$. But every pair of elements of S commutes; Proposition 1.2 now implies that $<S> = G$ is abelian, which is a contradiction. If P is a finite p-group and $|P : Z(P)| = p$, then $P/Z(P)$ is necessarily cyclic, contradicting the above. ∎

Recall that a proper subgroup H of a group G is said to be maximal if there is no proper subgroup of G that properly contains H. If G is a non-trivial finite group, then G has maximal subgroups, and every proper subgroup of G is contained in some maximal subgroup. We now show that maximal subgroups of finite p-groups are particularly well-behaved.

PROPOSITION 3. If P is a finite p-group, then every maximal subgroup of P is normal in P.

PROOF. We shall use induction on $|P|$. If $|P| = 1$, then P has no maximal subgroups; if $|P| = p$, then $P \cong \mathbf{Z_p}$ and the result holds. Thus, we can assume that $|P| > p$ and that the result is true for all p-groups of order less than $|P|$. Let M be a maximal subgroup of P, and let $Z = Z(P)$. Now $Z \trianglelefteq P$, so MZ is a subgroup of P which contains M; by the maximality of M, we have either $MZ = P$ or $MZ = M$. If $MZ = P$, then as M and Z are clearly contained in $N_G(M)$, we see that $M \triangleleft P$. Now suppose that $MZ = M$. Then $Z \trianglelefteq M$, and by the correspondence theorem we see that M/Z is a maximal subgroup of the p-group P/Z. Since $Z \neq 1$ by Theorem 1, we have $|P/Z| < |P|$, so by induction we have $M/Z \triangleleft P/Z$, which gives $M \triangleleft P$ by the correspondence theorem. ∎

COROLLARY 4. Any maximal subgroup of a finite p-group is of index p.

PROOF. Let P be a finite p-group and let M be a maximal subgroup of P; then $M \triangleleft P$ by Proposition 3. As P/M is a non-trivial p-group, it follows via Cauchy's theorem that P/M has a subgroup of order p, which must have the form L/M for some $M \triangleleft L \leqslant P$. But as M is maximal in P, we must have $L = P$, and consequently $|P : M| = p$. ∎

Our next task is to classify the finite abelian p-groups. We require the following lemma.

LEMMA 5. Every non-generator of a cyclic p-group P is a pth power in P.

PROOF. By Theorem 1.4, P has a unique subgroup Q of index p which is generated by x^p, where x is a generator of P. If $y \in Q$, then for some n we have $y = (x^p)^n = (x^n)^p$; hence every element of Q is a pth power in P. But by Corollary 4, Q must be the unique maximal subgroup of P and thus is exactly the set of non-generators of P, as the generators are precisely the elements that do not lie in any maximal subgroup. ∎

THEOREM 6. A finite abelian p-group is a direct product of cyclic p-groups.

PROOF. Let P be a finite abelian p-group. We shall use induction on $|P|$, so we may assume that $|P| > p$ and that the result is true for abelian p-groups of order less than $|P|$. Let Q be a maximal subgroup of P; then $|P/Q| = p$ by Corollary 4. By induction, we can write $Q = Q_1 \times \ldots \times Q_s$, where Q_i is cyclic of order p^{a_i} for each i; without loss of generality we can take $a_1 \geq \ldots \geq a_s \geq 1$.

Let x be an element of P which does not lie in Q. Since $|P/Q| = p$, we have $x^p \in Q$, and hence we have $x^p = y_1 \cdots y_s$, where $y_i \in Q_i$ for each i. If for some i and some $x_i \in Q_i$ we have $y_i = x_i^p$, then $(xx_i^{-1})^p = x^p x_i^{-p} = x^p y_i^{-1} = y_1 \cdots y_{i-1} \cdot 1 \cdot y_{i+1} \cdots y_s$ but $xx_i^{-1} \notin Q$. From this observation and Lemma 5, we conclude that there exists $x \in P - Q$ such that $x^p = y_1 \cdots y_s$, where each y_i is either a generator of Q_i or the identity. If $x^p = 1$, then P is the direct product of $<x>$ and the Q_i. Hence we can assume that $x^p \neq 1$, in which case $y_i \neq 1$ for some i; let $1 \leq j \leq s$ be minimal such that $y_j \neq 1$.

We now have $x^p = y_j \cdots y_s$. Since P is abelian, we see that the order of x^p is the least common multiple of the orders of $y_j, \ldots, y_s$, namely p^{a_j}; hence $|<x>| = p^{a_j+1}$. Let $\bar{Q} \leqslant Q$ be the direct product of all the Q_i except Q_j. We have $|\bar{Q}| = |Q|/|Q_j| = |Q|/p^{a_j}$. If we can show that $<x> \cap \bar{Q} = 1$, then $<x>\bar{Q}$ will be a direct product of order $p|Q| = |P|$, and hence P will be a direct product of cyclic p-groups.

Consider an element x^t of $<x>$. Since $x^p \in Q$ and $x^n \notin Q$ for $1 \leq n < p$, we see that $x^t \in Q$ only when $p \mid t$. Let $t = mp$ for some $0 \leq m < p^{a_j}$; then we have $x^t = (x^p)^m = y_j^m \cdots y_s^m$. Now $y_j^m \neq 1$, since $m < p^{a_j} = |<y_j>|$; thus the unique decomposition of x^t in $Q = Q_1 \times \ldots \times Q_s$ has a non-identity element in the coordinate associated to Q_j. This shows that x^t does not lie in $\bar{Q}$, since $\bar{Q}$ consists precisely of those elements of Q whose unique decomposition in the direct product Q has the identity in the Q_j-coordinate. Therefore, we have $<x> \cap \bar{Q} = 1$ as required. ∎

We now consider some of the ways in which knowledge about finite p-groups can be used to study arbitrary finite groups. Our next result is a standard tool in finite group theory and is of great importance in this regard.

FRATTINI ARGUMENT. Let G be a finite group, let $N \trianglelefteq G$, and let P be a Sylow subgroup of N. Then $G = N_G(P)N$.

PROOF. Let $g \in G$; then $gPg^{-1} \subseteq gNg^{-1} = N$. Hence P and gPg^{-1} are both Sylow subgroups of N, so by Sylow's theorem they are conjugate in N, and hence $n(gPg^{-1})n^{-1} = P$ for some $n \in N$. From this we see that $ng \in N_G(P)$, which allows us to conclude that $g \in N_G(P)n^{-1} \subseteq N_G(P)N$. ∎

THEOREM 7. The following statements about a finite group G are equivalent:

    (1) Every Sylow subgroup of G is normal in G.
    (2) G is the direct product of its Sylow subgroups.
    (3) Every maximal subgroup of G is normal in G.

A finite group satisfying these equivalent conditions is said to be *nilpotent*; clearly, finite abelian groups and finite p-groups are nilpotent. There is an equivalent definition of nilpotence which extends to infinite groups and which will be discussed in Section 11.

PROOF. We will prove the equivalence of the above statements by showing the circular implications $(1) \Rightarrow (2) \Rightarrow (3) \Rightarrow (1)$.

Suppose first that (1) holds. Let $p_1, \ldots, p_n$ be the distinct prime divisors of $|G|$, and for each $1 \le i \le n$ let P_i be a Sylow p_i-subgroup of G. By hypothesis, the P_i are normal subgroups of G. Using Lagrange's theorem and Propositions 1.7 and 3.12, we can show by induction that $P_1 \cdots P_i$ is a normal subgroup of order $|P_1| \cdots |P_i|$ for every i. This gives $G = P_1 \cdots P_n$. The same argument allows us to conclude that $|P_1 \cdots P_{i-1}P_{i+1} \cdots P_n| = |P_1| \cdots |P_{i-1}||P_{i+1}| \cdots |P_n|$, and hence that $P_i \cap P_1 \cdots P_{i-1}P_{i+1} \cdots P_n = 1$, for every i. It now follows that G is the direct product of the P_i. Therefore $(1) \Rightarrow (2)$.

Now suppose that (2) holds. Let M be a maximal subgroup of G, and let $P_1, \ldots, P_n$ be the Sylow subgroups of G. (The P_i are unique by hypothesis.) By Proposition 2.10, M is the direct product of the $M \cap P_i$. Since M is maximal in G, we see that there is exactly one j for which $M \cap P_j$ is maximal in P_j and that $M \cap P_i = P_i$ for all other i. As $M \cap P_j \lhd P_j$ by Proposition 3, we now see easily that $M \lhd G$. Therefore $(2) \Rightarrow (3)$.

Finally, suppose that (3) holds. Let P be a Sylow subgroup of G which is not normal in G. Then we have $P \le N_G(P) < G$, so there is some maximal subgroup M of G such that $P \le N_G(P) \le M$. As P is a Sylow subgroup of M, and $M \lhd G$ by hypothesis, the Frattini argument now gives $G = N_G(P)M \le M$, which is a contradiction. Therefore $(3) \Rightarrow (1)$. ∎

As a consequence of this result and our previous work, we can now deduce a result that is often called the "basis theorem:"

COROLLARY 8. A finite abelian group is a direct product of cyclic p-groups.

PROOF. By Theorem 7, any finite abelian group is the direct product of its Sylow subgroups; but by Theorem 6, each of those Sylow subgroups is a direct product of cyclic p-groups. ∎

We close this section with a discussion of p-groups of small order. Clearly, any group of order p is isomorphic with $\mathbf{Z_p}$. By Theorem 1 and Lemma 2, we see that any group of order p^2 is abelian and hence by Theorem 6 is isomorphic with either $\mathbf{Z_{p^2}}$ or $\mathbf{Z_p} \times \mathbf{Z_p}$. Theorem 6 also shows that any abelian group of order p^3 is isomorphic with one of $\mathbf{Z_{p^3}}, \mathbf{Z_{p^2}} \times \mathbf{Z_p}$, or $\mathbf{Z_p} \times \mathbf{Z_p} \times \mathbf{Z_p}$.

We now wish to classify the non-abelian groups of order p^3. We see from Theorem 1 and Lemma 2 that the center of such a group must have order p. We see also from Proposition 3 and Corollary 4 that a maximal subgroup of such a group is a normal subgroup of order p^2. For now we consider the case where p is odd. We will need the following rather technical lemma.

LEMMA 9. Let G be a group and let $x, y \in G$. If $[x, y] \in Z(G)$, then $[x^n, y] = [x, y]^n$ and $x^n y^n = (xy)^n [x, y]^{\frac{n(n-1)}{2}}$ for any $n \in \mathbb{N}$.

PROOF. Consider the first statement; we shall use induction on n. Suppose that the statement is true for $n \in \mathbb{N}$. Then we have

$$[x^{n+1}, y] = xx^n yx^{-n}(y^{-1}y)x^{-1}y^{-1} = x[x^n, y]yx^{-1}y^{-1}$$
$$= x[x, y]^n yx^{-1}y^{-1} = [x, y]^{n+1}$$

via the induction hypothesis.

Now consider the second statement; again we shall use induction on n. Assume that the statement holds true for $n \in \mathbb{N}$. Then

$$x^{n+1}y^{n+1} = xx^n y(yx^n)^{-1}(yx^n)y^n = x[x^n, y]yx^n y^n$$
$$= x[x, y]^n y(xy)^n [x, y]^{\frac{n(n-1)}{2}}$$
$$= (xy)^{n+1}[x, y]^{\frac{(n+1)n}{2}}$$

via the first statement and the induction hypothesis. ∎

PROPOSITION 10. If p is an odd prime, then there is exactly one isomorphism class of non-abelian groups of order p^3 having elements of order p^2.

PROOF. Let P be a non-abelian group of order p^3 having an element of order p^2. We wish to show that we can find elements x and y of P such that x has order p, y has order p^2, and $x \notin <y>$. In this case, we would have $P = <y> \rtimes <x>$, since the subgroup $<y>$ is maximal and therefore normal.

Choose an element y of order p^2, and choose some $x \notin <y>$. Assume that $x^p \neq 1$, for if not then we have found the desired elements x and y. We know that $|Z(P)| = p$, and by Theorem 1 we have $<y> \cap Z(P) \neq 1$; it follows that $Z(P)$ is a subgroup of $<y>$ having order p, which by Theorem 1.4 forces $Z(P) = <y^p>$. The group $P/Z(P)$ has order p^2 and, by Lemma 2, is not cyclic; hence it has exponent p. In particular, since $(xZ(P))^p = Z(P)$, we see that x^p is a non-trivial element of $Z(P)$, and so we must have $x^p = y^{kp}$ for some $1 \leq k < p$. Replace y by y^{-k}, so that $x^p = y^{-p}$; it is still true that y has order p^2 and that $x \notin <y>$. Now since P is non-abelian but $P/Z(P)$ is abelian, and since $Z(P)$ is simple, it follows from Proposition 2.6 that $P' = Z(P)$. In particular, we have $[x, y] \in Z(P)$, and by Lemma 9 we have $(xy)^p = x^p y^p [x, y]^{\frac{p(p-1)}{2}} = 1$. (Observe that 2 divides $p-1$ since p is odd.) Replace x by xy; we still have $x \notin <y>$, and hence x and y are the desired elements.

We have shown so far that any non-abelian group of order p^3 having an element of order p^2 can be written as a semidirect product of $\mathbf{Z_{p^2}}$ by $\mathbf{Z_p}$. From Proposition 2.1, we know that $\text{Aut}(\mathbf{Z_{p^2}})$ is an abelian group of order $p(p-1)$. Sylow's theorem now implies that $\text{Aut}(\mathbf{Z_{p^2}})$ has a unique subgroup of order p. Thus there exists a monomorphism from $\mathbf{Z_p}$ to $\text{Aut}(\mathbf{Z_{p^2}})$, and any two such maps must have the same image. The existence and uniqueness of a semidirect product of $\mathbf{Z_{p^2}}$ by $\mathbf{Z_p}$ now follows from Proposition 2.11. ∎

PROPOSITION 11. If p is an odd prime, then there is exactly one isomorphism class of non-abelian groups of order p^3 having no elements of order p^2.

PROOF. We easily see that any non-abelian group of order p^3 having no elements of order p^2 can be expressed as the semidirect product of a maximal subgroup by the subgroup generated by an element not lying in that maximal subgroup. Therefore, it suffices to consider semidirect products of $\mathbf{Z_p} \times \mathbf{Z_p}$ by $\mathbf{Z_p}$. Proposition 4.1 gives $\mathrm{Aut}(\mathbf{Z_p} \times \mathbf{Z_p}) \cong \mathrm{GL}(2,p)$, where if $\mathbf{Z_p} \times \mathbf{Z_p} = <u,v>$, then $\left(\begin{smallmatrix} a & b \\ c & d \end{smallmatrix}\right) \in \mathrm{GL}(2,p)$ corresponds to the automorphism sending u to $u^a v^c$ and v to $u^b v^d$. We have $|\mathrm{Aut}(\mathbf{Z_p} \times \mathbf{Z_p})| = p(p-1)^2(p+1)$ by Proposition 4.2, and hence all subgroups of order p of $\mathrm{Aut}(\mathbf{Z_p} \times \mathbf{Z_p})$ are conjugate by Sylow's theorem. Let ψ and τ be monomorphisms from $\mathbf{Z_p}$ to $\mathrm{Aut}(\mathbf{Z_p} \times \mathbf{Z_p})$. (Such monomorphisms exist; for instance, consider the map sending a generator of $\mathbf{Z_p}$ to the automorphism corresponding to $\left(\begin{smallmatrix} 1 & 1 \\ 0 & 1 \end{smallmatrix}\right)$.) Then there is some $f \in \mathrm{Aut}(\mathbf{Z_p} \times \mathbf{Z_p})$ such that $\tau(\mathbf{Z_p}) = f\psi(\mathbf{Z_p})f^{-1} = (\hat{f} \circ \psi)(\mathbf{Z_p})$, where $\hat{f}$ is the inner automorphism of $\mathrm{Aut}(\mathbf{Z_p} \times \mathbf{Z_p})$ induced by f. We now have $(\mathbf{Z_p} \times \mathbf{Z_p}) \rtimes_\tau \mathbf{Z_p} \cong (\mathbf{Z_p} \times \mathbf{Z_p}) \rtimes_{\hat{f} \circ \psi} \mathbf{Z_p} \cong (\mathbf{Z_p} \times \mathbf{Z_p}) \rtimes_\psi \mathbf{Z_p}$ by Propositions 2.11 and 2.12. ∎

We now discuss the non-abelian groups of order 8. By Exercise 1.2, any such group P must have an element of order 4. If P has elements x and y of respective orders 2 and 4 such that $x \notin <y>$, then by imitating the proof of Proposition 10 we conclude that $P \cong \mathbf{Z_4} \rtimes_\varphi \mathbf{Z_2} \cong D_8$, where φ is the unique monomorphism from $\mathbf{Z_2}$ to $\mathrm{Aut}(\mathbf{Z_4}) \cong \mathbf{Z_2}$. This will happen unless P has exactly one element of order 2, a possibility we must now investigate.

Suppose that t is the unique element of order 2 of P. Let $y \in P$ be an element of order 4, let $Q = <y> \lhd P$, and let $x \in P$ be such that $P/Q = <xQ>$. Since x cannot be of order 2, we must have $x^2 = t = y^2$. We now see that each element of P can be written uniquely as $x^a y^b$ for some $0 \le a \le 3$ and $0 \le b \le 1$. It is not hard to show that we must have $yx = x^3 y$; this completely specifies the group operation of P. It now follows that if such a group exists, then it is uniquely determined up to isomorphism. To show existence, we consider the subgroup of $\mathrm{SL}(2,3)$ generated by $x = \left(\begin{smallmatrix} 1 & 1 \\ 1 & 2 \end{smallmatrix}\right)$ and $y = \left(\begin{smallmatrix} 0 & 2 \\ 1 & 0 \end{smallmatrix}\right)$, which is non-abelian of order 8 and which has a unique element $\left(\begin{smallmatrix} 2 & 0 \\ 0 & 2 \end{smallmatrix}\right)$ of order 2. (Here we are thinking of the field of 3 elements as being the set $\{0, 1, 2\}$ with arithmetic performed modulo 3.) This group is called the *quaternion group*. We gave an alternate description of the quaternion group in Exercise 2.11, where it was shown that this group (unlike all other groups of order p^3) cannot be written non-trivially as a semidirect product.

EXERCISES

Throughout these exercises, p denotes a prime.

1. Show that if P is a non-cyclic finite p-group, then P has a normal subgroup N such that $P/N \cong \mathbf{Z_p} \times \mathbf{Z_p}$.

2. Let P be a group of order p^n. Show that P has a normal subgroup N_a of order p^a for every $0 \le a \le n$, and that these subgroups can be chosen so that N_a is contained in N_b whenever $a \le b$.

3. Let $G = \mathrm{GL}(n, p)$, and let P be a Sylow p-subgroup of G. What is the order of $Z(P)$? What is the order of $Z(P/Z(P))$? If we let $Z_2(P) \le P$ be such that $Z_2(P)/Z(P) = Z(P/Z(P))$ and continue in this way, what happens?

4. Let U be the subgroup of $\mathrm{GL}(n, p)$ consisting of the upper unitriangular matrices, and let Q be the subgroup of U consisting of all matrices whose (i, j)-entry is zero whenever $1 < i < j < n$. Determine $Z(Q)$, and show that $Q/Z(Q)$ is abelian.

5. Show that subgroups and quotient groups of finite nilpotent groups are nilpotent, and that direct products of finite nilpotent groups are nilpotent.

6. Let U be the subgroup of $\mathrm{GL}(3, p)$ consisting of the upper unitriangular matrices. Show that if p is an odd prime, then U is a non-abelian group of order p^3 having no elements of order p^2. If $p = 2$, with which group of order 8 is U isomorphic?

FURTHER EXERCISES

7. Show that a finite group G has a largest nilpotent normal subgroup, in the sense that it contains all nilpotent normal subgroups of G. (This subgroup is called the *Fitting subgroup* of G.)

The intersection of all maximal subgroups of a finite group G is called the *Frattini subgroup* of G and is denoted by $\Phi(G)$.

8. Show that $\Phi(G)$ is a nilpotent normal subgroup of G.

9. Show that $g \in \Phi(G)$ iff whenever $G = <S>$ and $g \in S$, then $G = <S - \{g\}>$.

10. Show that if P is a finite p-group, then $P/\Phi(P)$ is an elementary abelian p-group.

9. The Schur-Zassenhaus Theorem

This section is devoted to another of the basic results of finite group theory, the Schur-Zassenhaus theorem. We first need some terminology. A *complement* to a normal subgroup N of a group G is a subgroup H of G such that $G = N \rtimes H$. A subgroup H of a finite group G is called a *Hall subgroup* if $|H|$ and $|G : H|$ are coprime.

SCHUR-ZASSENHAUS THEOREM. *Any normal Hall subgroup of a finite group has a complement.*

PROOF. Let N be a normal Hall subgroup of a finite group G. If G has a subgroup K of order $n = |G : N|$, then we have $N \cap K = 1$ by Lagrange's theorem since n and $|N|$ are coprime, and hence we have $|NK| = |N||K| = |G|$ by Proposition 3.12, showing that K is a complement to N. Therefore, it suffices to show that G has a subgroup of order n. We shall assume by induction that any finite group of order less than $|G|$ which has a normal Hall subgroup also has a subgroup whose order is equal to the index of that normal Hall subgroup.

Let P be a Sylow subgroup of N. By the Frattini argument, we have $G = N_G(P)N$. Now $N_N(P) = N_G(P) \cap N \trianglelefteq N_G(P)$, and we have $G/N = N_G(P)N/N \cong N_G(P)/N_G(P) \cap N = N_G(P)/N_N(P)$ by the first isomorphism theorem. Thus $|N_G(P) : N_N(P)| = n$, and since $|N_N(P)|$ divides $|N|$, we conclude that $N_N(P)$ is a normal Hall subgroup of $N_G(P)$. If $N_G(P) < G$, then by induction $N_G(P)$, and hence G, has a subgroup of order n. Therefore, we assume that $N_G(P) = G$, or equivalently that $P \trianglelefteq G$.

Suppose that $P \vartriangleleft N$. By the correspondence theorem, we have $N/P \trianglelefteq G/P$ and $|G/P : N/P| = |G : N| = n$. Since $|N/P|$ divides $|N|$ and $|G/P| < |G|$, by induction G/P has a subgroup of order n; this subgroup must be of the form L/P where $P \vartriangleleft L \leqslant G$. Now $|L \cap N|$ divides both $|L| = n|P|$ and $|N|$, which since n and $|N|$ are coprime forces $|L \cap N| \leq |P|$. But $P \leqslant L \cap N$; hence $L \cap N = P$, and in particular $L < G$. As $|P|$ and $|L/P| = n$ are coprime, we see by induction that L, and hence G, has a subgroup of order n. We now assume that $N = P$.

Suppose that N is non-abelian. Let $Z = Z(N)$; then $1 < Z \vartriangleleft N$ by Theorem 8.1 since N is a p-group, and since the center of a group is a characteristic subgroup, it follows from Exercise 2.4 that $Z \vartriangleleft G$. By the correspondence theorem, G/Z has a normal subgroup N/Z

of index n. Thus by induction, G/Z has a subgroup of order n of the form L/Z where $Z \lhd L \leqslant G$. Arguing as in the previous paragraph, we find that $L \cap N = Z$ and in particular that $L < G$. Here $|Z|$ and $|L/Z| = n$ are coprime, so by induction L, and hence G, has a subgroup of order n.

It now suffices to prove that if A is an abelian normal Hall subgroup of G, then A has a complement in G. *Within the abelian group A, we shall adopt additive notation, but we will retain multiplicative notation when considering A as a subgroup of G.* (Our new choices of notation are due to the connections of what follows with the cohomology of groups; these connections will be discussed both after the proof and in the further exercises.)

Let $H = G/A$, and let $h \in H$; we view h as being a coset of A in G. If t and u are elements of G which are contained in the coset h, then we have $t^{-1}u \in A$ since $tA = uA = h$, and hence we have $txt^{-1} = uxu^{-1}$ for any $x \in A$ since A is abelian. Consequently, for $x \in A$ and $h \in H$, we can define ${}^h x \in A$ to be txt^{-1} for any $t \in h$. This gives an action of H on A; furthermore, this action has the additional property that ${}^h(x + y) = {}^h x + {}^h y$ for any $x, y \in A$, and ${}^h(-x) = -({}^h x)$ for any $x \in A$. (We can interpret this as giving a homomorphism from H to $\mathrm{Aut}(A)$.)

For each coset $h \in H$, we select some element $t_h \in h$; this gives an n-element set $\{t_h \mid h \in H\}$ which is a transversal of A in G. The fact that $t_{h_1 h_2}^{-1} A = (t_{h_1 h_2} A)^{-1} = (h_1 h_2)^{-1} = h_2^{-1} h_1^{-1}$ for any $h_1, h_2 \in H$ allows us to conclude that $t_{h_1} t_{h_2} t_{h_1 h_2}^{-1} \in A$. Thus we can use our choice of coset representatives to define a function $f : H \times H \to A$ by letting $f(h_1, h_2) \in A$ be such that $t_{h_1} t_{h_2} = f(h_1, h_2) t_{h_1 h_2}$. We have $t_{h_1}(t_{h_2} t_{h_3}) = (t_{h_1} t_{h_2}) t_{h_3}$ for any $h_1, h_2, h_3 \in H$ by associativity in G, but we also have

$$t_{h_1}(t_{h_2} t_{h_3}) = t_{h_1} f(h_2, h_3) t_{h_2 h_3} = t_{h_1} f(h_2, h_3) t_{h_1}^{-1} t_{h_1} t_{h_2 h_3}$$
$$= {}^{h_1} f(h_2, h_3) f(h_1, h_2 h_3) t_{h_1 h_2 h_3},$$

and $(t_{h_1} t_{h_2}) t_{h_3} = f(h_1, h_2) t_{h_1 h_2} t_{h_3} = f(h_1, h_2) f(h_1 h_2, h_3) t_{h_1 h_2 h_3}$. We conclude that the function f satisfies the *cocycle identity*

$$ {}^{h_1} f(h_2, h_3) + f(h_1, h_2 h_3) = f(h_1, h_2) + f(h_1 h_2, h_3) $$

for any $h_1, h_2, h_3 \in H$.

Suppose that a set map $c\colon H \to A$ satisfies

$$f(h_1, h_2) = c(h_1 h_2) - c(h_1) - {}^{h_1} c(h_2)$$

for all $h_1, h_2 \in H$. Then we would have

$$
\begin{aligned}
c(h_1 h_2) t_{h_1 h_2} &= c(h_1) \, {}^{h_1} c(h_2) f(h_1 h_2) t_{h_1 h_2} \\
&= c(h_1) t_{h_1} c(h_2) t_{h_1}^{-1} t_{h_1} t_{h_2} \\
&= c(h_1) t_{h_1} c(h_2) t_{h_2}
\end{aligned}
$$

for all $h_1, h_2 \in H$. In this case, the map $\varphi\colon H \to G$ defined by $\varphi(h) = c(h) t_h$ would be a group monomorphism, and its image would be a subgroup of G of order $|H| = n$. (If $1 \neq h \in H$, then $t_h \notin A$ and hence $c(h) t_h \neq 1$, showing that $\ker \varphi$ is trivial.) Hence it suffices to show the existence of such a function c.

Define a function $e\colon H \to A$ by $e(h) = \sum_{k \in H} f(h, k)$ for $h \in H$. Using the cocycle identity, we have

$$
\begin{aligned}
n f(h_1, h_2) + e(h_1 h_2) &= \sum_{h_3 \in H} (f(h_1, h_2) + f(h_1 h_2, h_3)) \\
&= \sum_{h_3 \in H} ({}^{h_1} f(h_2, h_3) + f(h_1, h_2 h_3)) \\
&= {}^{h_1}\Big(\sum_{k \in H} f(h_2, k)\Big) + \sum_{k \in H} f(h_1, k) \\
&= {}^{h_1} e(h_2) + e(h_1)
\end{aligned}
$$

and hence $n f(h_1, h_2) = -e(h_1 h_2) + e(h_1) + {}^{h_1} e(h_2)$ for any $h_1, h_2 \in H$. Since A is abelian, the map sending each element of A to its nth power is an endomorphism of A, and as n and $|A|$ are coprime we find that this map is an automorphism. Therefore, each $x \in A$ has a unique preimage under the nth power map, which we denote by $\frac{1}{n} x$; observe that $\frac{1}{n}(x + y) = \frac{1}{n} x + \frac{1}{n} y$ for any $x, y \in A$, and $\frac{1}{n}(-x) = -\frac{1}{n} x$ for any $x \in A$. We now define $c\colon H \to A$ by $c(h) = -\frac{1}{n} e(h)$ for $h \in H$, and it follows that

$$f(h_1, h_2) = \tfrac{1}{n}(-e(h_1 h_2) + e(h_1) + {}^{h_1} e(h_2)) = c(h_1 h_2) - c(h_1) + {}^{h_1} c(h_2)$$

for any $h_1, h_2 \in H$ as required. $\blacksquare$

We now wish to fit some of the ideas developed in the proof of the Schur-Zassenhaus theorem into a more general framework.

Let H be an arbitrary group, and let A be an abelian group written additively. Suppose that we have an action of H as automorphisms of A, or equivalently a homomorphism from H to $\mathrm{Aut}(A)$. A function $f\colon H \times H \to A$ satisfying the cocycle identity given in the above proof is called a *2-cocycle* of the pair (H, A). The set of all 2-cocycles of (H, A) is denoted by $Z^2(H, A)$, and if we define $(f + g)(h_1, h_2) = f(h_1, h_2) + g(h_1, h_2)$ for $f, g \in Z^2(H, A)$ and $h_1, h_2 \in H$, then $Z^2(H, A)$ becomes an abelian group. A 2-cocycle is called a *2-coboundary* if there is a function $c\colon H \to A$ such that $f(h_1, h_2) = c(h_1) + h_1 c(h_2) - c(h_1 h_2)$ for all $h_1, h_2 \in H$. The set of 2-coboundaries of (H, A) is denoted by $B^2(H, A)$ and is a subgroup of $Z^2(H, A)$. The quotient group $Z^2(H, A)/B^2(H, A)$ is called the *second cohomology group* of (H, A) and is denoted by $H^2(H, A)$. (This terminology comes from algebraic topology. We shall define the nth cohomology group of (H, A) for any $n \in \mathbb{N}$ in the further exercises to Section 12.)

Now suppose that A is an abelian normal Hall subgroup of a finite group G. We showed during the proof of the Schur-Zassenhaus theorem that any 2-cocycle f of $(G/A, A)$ was a 2-coboundary, or equivalently that $H^2(G/A, A) = 0$. (The difference of sign between the definition of 2-coboundary given above and what was established in the proof is irrelevant.) Furthermore, we showed that this fact implied the existence of a complement to A in G. We attempt to make this connection between the second cohomology group of $(G/A, A)$ and the existence of complements to A in G more transparent in the further exercises below.

A stronger version of the Schur-Zassenhaus theorem asserts that if N is a normal Hall subgroup of a finite group G, then not only does N have a complement in G, but all such complements are conjugate in G. To prove this, one needs the concept of solvable groups, which we shall introduce in Section 11; see [22, pp. 246–8] for further details.

EXERCISES

1. Suppose that G is a finite group such that $G = N \rtimes H$, where H is abelian. Show that if $|N|$ and $|H|$ are relatively coprime, then all complements to N in G are conjugate.
2. Repeat Exercise 1 under the assumption that H is nilpotent rather than abelian.

Further Exercises

These exercises are a continuation of the further exercises to Section 2. Let N and H be groups. A *factor pair* of N by H is a pair (f, φ) of set maps $f \colon H \times H \to N$ and $\varphi \colon H \to \operatorname{Aut}(N)$ satisfying properties **1**, **2**, and **3** listed on page 27. Let $\mathcal{E}$ be the set of extensions of N by H, and let $\mathcal{F}$ be the set of factor pairs of N by H. In what follows, we will always use "extension" to mean an element of $\mathcal{E}$, and "factor pair" to mean an element of $\mathcal{F}$.

3. Let (f, φ) be a factor pair, and define

$$(x, \alpha) \cdot (y, \beta) = (x\varphi(\alpha)(y)f(\alpha, \beta), \alpha\beta)$$

for $(x, \alpha), (y, \beta) \in N \times H$. Show that this gives a group structure on $N \times H$; call this group $E_{f,\varphi}$. Show further that $(E_{f,\varphi}, i, \pi)$ is an extension, where $i(x) = (x, 1)$ and $\pi(x, \alpha) = \alpha$, and that (f, φ) is the factor pair arising from some normalized section of $E_{f,\varphi}$. (Observe that this construction generalizes the notion of external semidirect product.)

We have seen in Exercise 2.13 that an extension gives rise to a factor pair via a choice of normalized section, and we have just given an explicit construction of an extension from a given factor pair. We view these processes as giving maps between $\mathcal{E}$ and $\mathcal{F}$, and we now investigate the relationship between these maps. We must first consider the relation between factor pairs arising from different normalized sections of the same extension.

4. (cont.) Suppose that t and u are normalized sections of an extension E, and let (f, φ) and (g, ρ) be the factor pairs arising from t and u, respectively. Let $c \colon H \to N$ be the set map such that $u(\alpha) = c(\alpha)t(\alpha)$ for every $\alpha \in H$. Show that the following properties hold:

   4 $\rho(\alpha) = \Psi(c(\alpha))\varphi(\alpha)$ for $\alpha \in H$, where $\Psi(c(\alpha))$ is the inner automorphism of N coresponding to $c(\alpha)$.
   5 $g(\alpha, \beta) = c(\alpha)\varphi(\alpha)(c(\beta))f(\alpha, \beta)c(\alpha\beta)^{-1}$ for $\alpha, \beta \in H$.

The above exercise motivates the following definition: We say that two factor pairs (f, φ) and (g, ρ) are *equivalent* if there is a map $c \colon N \to H$ such that properties **4** and **5** hold. (Verify that this is an equivalence relation on $\mathcal{F}$.) Let $\overline{\mathcal{F}}$ denote the set of equivalence classes of factor pairs; we will use $[f, \varphi]$ to denote the class of the factor pair (f, φ). We have a well-defined map from $\mathcal{E}$ to $\overline{\mathcal{F}}$ which sends an extension to the class of a factor pair arising from any normalized section. We must now consider what happens when we pass from $\overline{\mathcal{F}}$ back to $\mathcal{E}$ via the construction in Exercise 3. Here we will need to recall the exact definition of an extension.

5. (cont.) Let (f, φ) and (g, ρ) be factor pairs, with $[f, \varphi] = [g, \rho]$. Let $i \colon N \to E_{f,\varphi}$ and $j \colon N \to E_{g,\rho}$ be the natural inclusions (of N into the underlying set $N \times H$), and let $\pi \colon E_{f,\varphi} \to H$ and $\tau \colon E_{g,\rho} \to H$ be the natural projections (of the underlying set $N \times H$ onto H). Show that there is an isomorphism $\xi \colon E_{f,\varphi} \to E_{g,\rho}$ such that $\xi \circ i = j$ and $\tau \circ \xi = \pi$.

Motivated by the above exercise, we say that two extensions (E, i, π) and (F, j, τ) are *equivalent* if there is an isomorphism $\xi \colon E \to F$ such that $\xi \circ i = j$ and $\tau \circ \xi = \pi$. (Verify that this gives an equivalence relation on $\mathcal{E}$.) We let $\overline{\mathcal{E}}$ denote the set of equivalence classes of extensions, and we let $[E]$ denote the class of an extension E. In this context, Exercise 5 asserts that there is a well-defined map from $\overline{\mathcal{F}}$ to $\overline{\mathcal{E}}$, sending $[f, \varphi]$ to $[E_{f,\varphi}]$.

6. (cont.) If p is an odd prime, show that $\mathbf{Z}_{p^2}$ can be realized in $p - 1$ nonequivalent ways as an extension of $\mathbf{Z_p}$ by $\mathbf{Z_p}$.

7. (cont.) We have already obtained a map from $\mathcal{E}$ to $\overline{\mathcal{F}}$, sending an extension E to the class of the factor pair arising from any normalized section of E. Show that this map induces a map from $\overline{\mathcal{E}}$ to $\overline{\mathcal{F}}$.

8. (cont.) Show that the map from $\overline{\mathcal{E}}$ to $\overline{\mathcal{F}}$ obtained in Exercise 7 is inverse to the map from $\overline{\mathcal{F}}$ to $\overline{\mathcal{E}}$ sending $[f, \varphi]$ to $[E_{f,\varphi}]$. Conclude that there is a bijective correspondence between the set of equivalence classes of extensions and the set of equivalence classes of factor pairs.

Exercise 8 implies that in order to study extensions up to equivalence, it suffices to study equivalence classes of factor pairs. The next two exercises give a slight refinement of the correspondence just obtained.

9. (cont.) Let $\eta \colon \operatorname{Aut}(N) \to \operatorname{Out}(N)$ be the natural map. Show that if (f, φ) and (g, ρ) are any two factor pairs arising from an extension E, then $\eta \circ \varphi = \eta \circ \rho$, and this map from H to $\operatorname{Out}(N)$ (which we denote by ψ_E) is a homomorphism. Conclude that there is a well-defined map from $\mathcal{E}$ to the set of homomorphisms from H to $\operatorname{Out}(N)$, sending E to ψ_E.

10. (cont.) Let $\psi \colon H \to \operatorname{Out}(N)$ be a given homomorphism. Show that there is a bijective correspondence between the set of classes $[E]$ of $\overline{\mathcal{E}}$ for which $\psi_E = \psi$ and the set of classes $[f, \varphi]$ of $\overline{\mathcal{F}}$ for which $\eta \circ \varphi = \psi$, where $\eta \colon \operatorname{Aut}(N) \to \operatorname{Out}(N)$ is the natural map.

We now consider the case where the group N is abelian; we write A instead of N, and we will use additive notation for A. Observe that $\operatorname{Out}(A) = \operatorname{Aut}(A)$. We fix a homomorphism $\varphi \colon H \to \operatorname{Aut}(A)$, and we write xa in lieu of $\varphi(x)(a)$ for $x \in H$ and $a \in A$. We would like to study those equivalence classes $[E]$ of extensions of N by H for which $\psi_E = \varphi$; we say that such extensions *respect* the action of H on A. By Exercise 10, it suffices to study equivalence classes $[f, \varphi]$ of factor pairs of A by H. We suppress φ in our notation, so that we are studying functions $f \colon H \times H \to A$ such that $f(x, 1) = f(1, x) = 0$ for all $x \in H$ and which in addition satisfy

$$f(x, y) + f(xy, z) = xf(y, z) + f(x, yz)$$

for all $x, y, z \in H$, with two such functions f and g being equivalent if there is a map $c \colon H \to A$ such that $c(1) = 0$ and

$$g(x, y) = f(x, y) + c(x) + xc(y) - c(xy)$$

for all $x, y \in H$. As discussed in the section, the set $H^2(H, A)$ of equivalence classes of such functions forms an abelian group that is called the second cohomology group of (H, A). It now follows from Exercise 10 that there is a bijective correpondence between the group $H^2(H, A)$ and the set of equivalence classes of extensions of A by H which respect the action of H on A. In particular, if $H^2(H, A) = 0$, then every extension of A by H is split.

11. (cont.) Suppose that H and A are both finite. Show that the order of each element of $H^2(H, A)$ divides both $|H|$ and the exponent of A. (This implies that $H^2(G/A, A) = 0$ when A is an abelian normal Hall subgroup of a finite group G, which we established in proving the Schur-Zassenhaus theorem.)

4
Normal Structure

The theme of this chapter is the examination of a group G through the study of descending series of subgroups of G in which each term is either normal in G or at least normal in the previous term. These series allow us to consider the "normal structure" of groups, an area of study that is fundamental to group theory. In Section 10 we discuss composition series and chief series, while in Section 11 our attention turns to the derived series and to central series, and hence to the concepts of solvability and nilpotence.

10. Composition Series

We say that a series of subgroups $G = G_0 > G_1 > \ldots > G_r = 1$ of a group G is a *composition series* of G if $G_{i+1} \lhd G_i$ for every i and if each *successive quotient* G_i/G_{i+1} is simple. The above composition series is said to have length r. The successive quotients of a composition series are called the *composition factors* of the series. More generally, a group is said to be a *composition factor* of G if it is isomorphic with one of the composition factors in some composition series of G.

For example, consider the group Σ_5. It has a normal subgroup A_5 that is simple by Theorem 7.4. Since $\Sigma_5/A_5 \cong \mathbf{Z_2}$ is also simple, we see that $\Sigma_5 > A_5 > 1$ is a composition series of Σ_5. In fact, this is the only composition series of Σ_5, for we observed in the proof of Theorem 7.5 that the only non-trivial proper normal subgroup of Σ_5 is A_5.

We say that N is a *maximal normal subgroup* of a group G if $N \lhd G$ and if there is no proper normal subgroup of G that properly contains N. Any non-trivial finite group has maximal normal subgroups. We see using the correspondence theorem that N is a maximal normal subgroup of G iff G/N is simple. A maximal subgroup that is also normal is clearly a maximal normal subgroup (and must be of prime index), but a maximal normal subgroup need not be a maximal subgroup, for it could be properly contained in a proper subgroup that is not normal. For example, the group $A_5 \times \mathbf{Z_2}$ has $1 \times \mathbf{Z_2}$ as a maximal normal subgroup that is not maximal.

PROPOSITION 1. Finite groups have composition series.

PROOF. Let G be a finite group. We use induction on $|G|$. If G is simple, then $G > 1$ is a composition series of G, and otherwise G has some maximal normal subgroup G_1, which has a composition series $G_1 > G_2 > \ldots > G_r = 1$ by induction. Since G/G_1 is simple, it now follows that $G = G_0 > G_1 > \ldots > G_r = 1$ is a composition series of G. ∎

Infinite groups need not have composition series. For example, using Theorem 1.5 we see that every non-trivial subgroup of the infinite cyclic group $\mathbf{Z}$ is isomorphic with $\mathbf{Z}$; as $\mathbf{Z}$ is not simple, we conclude that $\mathbf{Z}$ has no simple subgroups and hence that we cannot construct a composition series of $\mathbf{Z}$, since the last non-trivial term of such a series must be a simple subgroup of $\mathbf{Z}$.

LEMMA 2. Let G be a group that has a composition series, and let $N \trianglelefteq G$. Then N has a composition series.

PROOF. Let $G = G_0 > G_1 > \ldots > G_r = 1$ be a composition series of G. Let $N_i = N \cap G_i$ for each i, so that we have a series $N = N_0 \geqslant N_1 \geqslant \ldots \geqslant N_r = 1$ of subgroups of N. Fix some i. We easily see that $N_{i+1} \trianglelefteq N_i$, and since $N \cap G_{i+1} = (N \cap G_i) \cap G_{i+1}$ we have $N_i/N_{i+1} = N \cap G_i/N \cap G_{i+1} \cong (N \cap G_i)G_{i+1}/G_{i+1}$ by the first isomorphism theorem. Let $\eta \colon G_i \to G_i/G_{i+1}$ be the natural map; then $(N \cap G_i)G_{i+1}/G_{i+1} = \eta(N \cap G_i) \trianglelefteq \eta(G_i) = G_i/G_{i+1}$, and hence N_i/N_{i+1} is isomorphic with a normal subgroup of the simple group G_i/G_{i+1}. Therefore, either $N_i = N_{i+1}$, or $N_i/N_{i+1} \cong G_i/G_{i+1}$ is simple; thus we obtain a composition series of N by deleting from the series $N = N_0 \geqslant N_1 \geqslant \ldots \geqslant N_r = 1$ any repetitions that may occur. ∎

Let $G = G_0 > G_1 > \ldots > G_r = 1$ be a composition series, and suppose that $G = H_0 > H_1 > \ldots > H_r = 1$ is another composition series of the same length r. We say that these series are *equivalent* if there is some $\rho \in \Sigma_r$ such that $G_{i-1}/G_i \cong H_{\rho(i)-1}/H_{\rho(i)}$ for every i. For example, let $G = <x> \cong \mathbf{Z_6}$, let $G_1 = <x^2>$, let $H_1 = <x^3>$, and consider the two composition series $G > G_1 > 1$ and $G > H_1 > 1$; these are equivalent, as $G/G_1 \cong H_1/1 \cong \mathbf{Z_2}$ and $G_1/1 \cong G/H_1 \cong \mathbf{Z_3}$. (Here we take $\rho = (1\ 2) \in \Sigma_2$.)

Our next result asserts that, up to equivalence, a group has at most one composition series. Consequently, a group having a composition series has a well-defined collection of composition factors, and an understanding of these factors gives a framework for gaining further information about the group. Since composition factors are simple groups, in studying arbitrary finite groups we would like to have comprehensive knowledge about the finite simple groups. The classification of all finite simple groups was completed in 1980 after two decades of concentrated work by a number of specialists, and it is generally regarded as one of the crown jewels of twentieth-century mathematics.

JORDAN-HÖLDER THEOREM. Suppose that G is a group that has a composition series. Then any two composition series of G have the same length and are equivalent.

PROOF. Suppose that $G = G_0 > G_1 > \ldots > G_r = 1$ and $G = H_0 > H_1 > \ldots > H_s = 1$ are two composition series of G. We use induction on r, the length of one of the composition series. If $r = 1$, then G is simple, and clearly in this case $G > 1$ is the only composition series of G. Now let $r > 1$, and assume by induction that the result holds for any group having some composition series of length less than r. If $G_1 = H_1$, then G_1 has two composition series of respective lengths $r-1$ and $s-1$, and so by induction we see that $r = s$ and that the two composition series of G_1, and hence the two composition series of G, are equivalent.

We now assume that $G_1 \neq H_1$. As $G_1 \trianglelefteq G$ and $H_1 \trianglelefteq G$, we have $G_1 H_1 \trianglelefteq G$ by Proposition 1.7. But G/G_1 is simple, so we cannot have $G_1 \leqslant H_1$, and hence we must have $H_1 < G_1 H_1$, which since G/H_1 is simple forces $G_1 H_1 = G$. Let $K = G_1 \cap H_1 \trianglelefteq G$, and observe that we have $G/G_1 \cong H_1/K$ and $G/H_1 \cong G_1/K$ by the first isomorphism theorem. (In particular, G_1/K and H_1/K are simple.) We have a composition series $K = K_0 > K_1 > \ldots > K_t = 1$ of K by Lemma 2.

We now have two composition series $G_1 > G_2 > \ldots > G_r = 1$ and $G_1 > K > K_1 > K_2 > \ldots > K_t = 1$ of G_1. These are of lengths $r - 1$ and $t+1$, respectively; by induction, we see that $t = r-2$ and that the series are equivalent. Similarly, we now have two composition series $H_1 > H_2 > \ldots > H_s = 1$ and $H_1 > K > K_1 > K_2 > \ldots > K_{r-2} = 1$ of H_1. As these have respective lengths $s - 1$ and $r - 1$, by induction we see that $r = s$ and that the series are equivalent. Finally, we observe that, because of the isomorphisms derived in the previous paragraph, the composition series $G = G_0 > G_1 > K > \ldots > K_{r-2} = 1$ and $G = H_0 > H_1 > K > \ldots > K_{r-2} = 1$ are equivalent. It now follows that our two initial composition series of G are equivalent. ∎

Composition series are but one of many types of series of subgroups that play important roles in group theory, and so we now wish to introduce some general terminology. A series of subgroups $G = G_0 \geqslant G_1 \geqslant \ldots \geqslant G_r = 1$ of a group G is called a *subnormal series* of G if $G_{i+1} \trianglelefteq G_i$ for every i; a subnormal series is called a *normal series* if $G_i \trianglelefteq G$ for every i. (Some authors use the terminology "normal series" for what we call a subnormal series, and use "invariant series" for what we call a normal series.) Composition series are examples of subnormal series, but subnormal series can have successive quotients that are trivial or that are non-trivial but not simple. Two subnormal series of the same length are said to be *equivalent* if they satisfy the condition stated as the definition of equivalence of composition series. Any subnormal series that is obtained from a given subnormal series by interposing additional terms is called a *refinement* of the given subnormal series, and such a refinement is said to be *proper* if at least one of the interposed terms was not already present in the series. With this terminology, a composition series is a subnormal series that has no repeated terms and does not admit a proper refinement.

A *chief series* of G is a normal series of G with no repeated terms and with the additional property that no normal subgroup of G is contained properly between any two terms of the series. Notice the analogy: A composition series is a subnormal series having no subnormal series as a proper refinement, and a chief series is a normal series having no normal series as a proper refinement. The *chief factors* of a chief series are the succesive quotients, while a group is said to be a *chief factor* of G if it is isomorphic with a chief factor in some chief series of G. The analogue for chief series of the Jordan-Hölder theorem, namely that any two chief series of a group have the same length and are equivalent, is true, and the proof is virtually identical to the proof given in the case of composition series. Both of these results are special cases of a more general result, the Jordan-Hölder theorem for groups with operators, which we shall not discuss here. (See [26, Section 2.3].)

We say that N is a *minimal normal subgroup* of a group G if $1 \neq N \trianglelefteq G$ and if there is no non-trivial normal subgroup of G that is properly contained in N. Any non-trivial finite group has minimal normal subgroups, and a simple group has a unique minimal normal subgroup, namely itself.

PROPOSITION 3. Finite groups have chief series.

PROOF. Let G be a finite group. We shall use induction on $|G|$. If G is simple, then $G > 1$ is a chief series for G; otherwise, G has a proper minimal normal subgroup N. By induction, G/N has a chief series, which by the correspondence theorem has the form $G/N = G_0/N > G_1/N > \ldots > G_r/N = 1$ where, for each i, $G_i \trianglelefteq G$ and no normal subgroup of G lies properly between G_{i-1} and G_i. We now see that $G = G_0 > G_1 > \ldots > G_r = N > 1$ is a chief series for G, since N is a minimal normal subgroup. ∎

By definition, the composition factors of finite groups are simple groups. We will complete this section by determining the nature of chief factors of finite groups. Our first step is to rephrase this problem in different terms.

LEMMA 4. Let G be a group having a chief series. Then every chief factor of G is a minimal normal subgroup of a quotient group of G.

PROOF. If $G = G_0 > G_1 > \ldots > G_r = 1$ is a chief series of G, then it follows from the correspondence theorem that the chief factor G_i/G_{i+1} is a minimal normal subgroup of G/G_{i+1}. ∎

THEOREM 5. A minimal normal subgroup of a finite group is a direct product of mutually isomorphic simple groups.

PROOF. Let G be a finite group and let N be a minimal normal subgroup of G. Let N_1 be a maximal normal subgroup of N, so that N/N_1 is simple. Let $N_1, N_2, \ldots, N_r$ be the conjugates of N_1 in G; since $N \trianglelefteq G$, each N_i is a maximal normal subgroup of N. If $N_i = x N_1 x^{-1}$ for some $x \in G$, then the map from N/N_1 to N/N_i sending $g N_1$ to $x g x^{-1} N_i$ is a well-defined isomorphism, showing that the groups N/N_i are mutually isomorphic. Now since the N_i are the distinct conjugates of N_1 in G, conjugation by $g \in G$ permutes the set $\{N_1, \ldots, N_r\}$, and therefore

$$g(N_1 \cap \ldots \cap N_r)g^{-1} = g N_1 g^{-1} \cap \ldots \cap g N_r g^{-1} = N_1 \cap \ldots \cap N_r.$$

Hence $N_1 \cap \ldots \cap N_r \trianglelefteq G$. But $N_1 \cap \ldots \cap N_r < N$, so by the minimality of N we must have $N_1 \cap \ldots \cap N_r = 1$.

We will show that, for each $1 \leq i \leq r$, the group $N/N_1 \cap \ldots \cap N_i$ is a direct product of groups isomorphic with N/N_1; the case $i = r$ completes the proof. We shall use induction on i. The case $i = 1$ is trivial, so we take $i > 1$ and assume that the result holds for $i - 1$. If $N_1 \cap \ldots \cap N_{i-1} \leqslant N_i$, then $N_1 \cap \ldots \cap N_i = N_1 \cap \ldots \cap N_{i-1}$ and hence there is nothing to prove. Thus, we assume that $N_1 \cap \ldots \cap N_{i-1} \nleqslant N_i$, in which case we have $N_i < (N_1 \cap \ldots \cap N_{i-1})N_i \trianglelefteq N$. We must have $(N_1 \cap \ldots \cap N_{i-1})N_i = N$, since N_i is a maximal normal subgroup of N. We now have

$$N/N_1 \cap \ldots \cap N_i = N_1 \cap \ldots \cap N_{i-1}/N_1 \cap \ldots \cap N_i \times N_i/N_1 \cap \ldots \cap N_i$$

by Lemma 2.7. But we have

$$N_1 \cap \ldots \cap N_{i-1}/N_1 \cap \ldots \cap N_{i-1} \cap N_i \cong (N_1 \cap \ldots \cap N_{i-1})N_i/N_i$$
$$= N/N_i \cong N/N_1$$

by the first isomorphism theorem, and similarly we have

$$N_i/N_1 \cap \ldots \cap N_{i-1} \cap N_i \cong (N_1 \cap \ldots \cap N_{i-1})N_i/N_1 \cap \ldots \cap N_{i-1}$$
$$= N/N_1 \cap \ldots \cap N_{i-1}.$$

What we wished to show now follows by induction. ∎

COROLLARY 6. A chief factor of a finite group is a direct product of mutually isomorphic simple groups.

PROOF. This follows directly from Lemma 4 and Theorem 5. ∎

EXERCISES

1. Show that an abelian group has a composition series iff it is finite.
2. Show that $GL(n, F)$ has a composition series iff F is finite.
3. Using the third isomorphism theorem (Exercise 2.14), prove the *Schreier refinement theorem*: Any two subnormal series of a given group have equivalent refinements.
4. (cont.) Use the Schreier refinement theorem to give an alternate proof of the Jordan-Hölder theorem.
5. Determine all composition series and all chief series of Σ_n for $n \geq 2$. (HINT: Use Exercise 3.8.)
6. Show that any group having a composition series has a chief series.
7. (cont.) Show that any chief factor of a group having a composition series is a direct product of mutually isomorphic simple groups.
8. Show that any finite group that has no proper non-trivial characteristic subgroups is a direct product of mutually isomorphic simple groups. Use this to give a new proof of Theorem 5.

11. Solvable Groups

We define a series $G^{(k)}$ of subgroups of a group G by setting $G^{(0)} = G$ and taking $G^{(k)}$ to be the derived group of $G^{(k-1)}$ for $k \in \mathbb{N}$. This series is called the *derived series* of G. Since $G^{(k+1)}$ is a characteristic subgroup of $G^{(k)}$ for each k by Lemma 2.5, we see via Lemma 2.4 that the derived series is a normal series of G. Moreover, we see from Proposition 2.6 that each successive quotient $G^{(k)}/G^{(k+1)}$ of the derived series is abelian.

We say that a group is *solvable* if its derived series terminates in the identity. (Authors from British Commonwealth countries often write "soluble" in lieu of "solvable" and hence "insoluble" in lieu of "non-solvable.") Since a group is abelian iff its derived group is trivial, we see that abelian groups are solvable. However, not all groups are solvable. For example, consider the group A_5. Since A_5 is non-abelian, its derived group must be a non-trivial normal subgroup of A_5. But A_5 is simple by Theorem 7.5, so the derived group of A_5 must be A_5, and hence we have $A_5^{(k)} = A_5$ for all k. (A group whose derived group is itself is said to be *perfect*; by this same argument, any non-abelian simple group is perfect, and in particular is not solvable. However, a perfect group need not be simple. For example, we established in Section 6 that if there is a non-trivial nth root of unity in F, then $\mathrm{SL}(n, F)$ is perfect but has non-trivial center, except when $n = 2$ and $|F| = 3$.)

Solvable groups are to be thought of as the opposite of simple groups, in that simple groups have very few normal subgroups while solvable groups are rife with them. If this is a valid mode of thinking, then there should be very few groups which are both simple and solvable. This is indeed the case:

PROPOSITION 1. A simple solvable group has prime order.

PROOF. Let G be a simple solvable group. Since G is solvable, we cannot have $G' = G$; as G is simple and $G' \trianglelefteq G$, this forces $G' = 1$, and so G is abelian. But every non-identity element of an abelian simple group must be a generator, and hence such a group must be finite and of prime order. ∎

We now give some alternate characterizations of solvability.

PROPOSITION 2. The following statements about a group G are equivalent:

(1) G is solvable.
(2) G has a normal series in which every successive quotient is abelian.
(3) G has a subnormal series in which every successive quotient is abelian.

PROOF. Clearly, we have (1) $\Rightarrow$ (2) $\Rightarrow$ (3), so we need only prove that (3) $\Rightarrow$ (1). Suppose that we have a subnormal series $G = G_0 \geqslant G_1 \geqslant \ldots \geqslant G_r = 1$ in which each successive quotient is abelian. To show that G is solvable, it suffices to show that $G^{(i)} \leqslant G_i$ for each i, as this gives $G^{(r)} \leqslant G_r = 1$. We shall use induction on i. Since G/G_1 is abelian, we have $G^{(1)} \leqslant G_1$ by Proposition 2.6. Now let $i > 1$ and assume by induction that $G^{(i-1)} \leqslant G_{i-1}$. Then $G^{(i)} = (G^{(i-1)})' \leqslant (G_{i-1})'$, and $(G_{i-1})' \leqslant G_i$ by Proposition 2.6 since G_{i-1}/G_i is abelian. ∎

We now consider the relationship between the solvability of a given group and that of related groups.

PROPOSITION 3.   (i) If G is solvable and $H \leqslant G$, then H is solvable.

(ii) If G is solvable and $N \trianglelefteq G$, then G/N is solvable.

(iii) If $N \trianglelefteq G$ and both N and G/N are solvable, then G is solvable.

(iv) If G and H are solvable, then $G \times H$ is solvable.

PROOF.   (i) This is clear, since $H^{(k)} \leqslant G^{(k)}$ for all k.

(ii) There is a normal series $G = G_0 \geqslant G_1 \geqslant \ldots \geqslant G_r = 1$ by Proposition 2 such that G_i/G_{i+1} is abelian for each i. Consider the series $G/N = G_0 N/N \geqslant G_1 N/N \geqslant \ldots \geqslant G_r N/N = 1$. Fix some i. Since $G_i \trianglelefteq G$ and $N \trianglelefteq G$, we have $G_i N \trianglelefteq G$ and hence $G_i N/N \trianglelefteq G/N$. Since $G_i N = G_i(G_{i+1}N)$, we have $(G_i N/N)/(G_{i+1} N/N) \cong G_i N/G_{i+1} N \cong G_i/G_i \cap G_{i+1} N$ by the first and second isomorphism theorems. We now see via the second isomorphism theorem that $G_i/G_i \cap G_{i+1} N$ is a quotient of the abelian group G_i/G_{i+1} and hence is abelian. Therefore, we have constructed a normal series of G/N having abelian successive quotients, and hence G/N is solvable by Proposition 2.

(iii) There are subnormal series $N = N_0 \geqslant N_1 \geqslant \ldots \geqslant N_r = 1$ and $G/N = G_0/N \geqslant G_1/N \geqslant \ldots \geqslant G_s/N = 1$ by Proposition 2 such that N_i/N_{i+1} and $(G_i/N)/(G_{i+1}/N) \cong G_i/G_{i+1}$ are abelian for each i. We now see that

$$G = G_0 \geqslant G_1 \geqslant \ldots \geqslant G_s = N = N_0 \geqslant N_1 \geqslant \ldots \geqslant N_r = 1$$

is a subnormal series of G having abelian successive quotients, and hence that G is solvable by Proposition 2.

(iv) Here $1 \times H \cong H$ is a solvable normal subgroup of $G \times H$, and $G \times H/1 \times H \cong G$ is also solvable, so it follows from part (iii) that $G \times H$ is solvable. ∎

We can now deduce a more concrete equivalent condition to solvability:

PROPOSITION 4. A group having a composition series is solvable iff all of its composition factors have prime order. (In particular, such groups are finite.)

PROOF. Let G be a group having a composition series. If all of the composition factors of G have prime order, then the composition series is a subnormal series having abelian successive quotients, and hence G is solvable by Proposition 2. Now suppose that G is solvable, and let H/K be a composition factor of G, where $K \lhd H \leqslant G$. Using parts (i) and (ii) of Proposition 3, we see that H/K is solvable. Therefore, H/K is a simple solvable group and hence has prime order by Proposition 1. ∎

We now see that there are groups, such as Σ_5, which are neither simple nor solvable, and that there are non-abelian groups, such as Σ_3 (which has the composition series $\Sigma_3 > A_3 > 1$), which are solvable. Proposition 4 also implies that an infinite abelian group, being solvable, cannot possess a composition series (which was Exercise 10.1).

COROLLARY 5. Finite p-groups are solvable.

PROOF. This follows from Proposition 4 and the fact that the composition factors of finite p-groups are finite simple p-groups and hence must be of prime order. ∎

We can extend the argument of Proposition 4 to give another characterization of solvability:

PROPOSITION 6. A group having a composition series is solvable iff all of its chief factors are elementary abelian.

(Any group having a composition series also has a chief series by Exercise 10.6. Alternately, a solvable group having a composition series is finite by Proposition 4 and hence has a chief series by Proposition 10.3.)

PROOF. Let G be a group having a composition series. If all of the chief factors of G are elementary abelian, then by refining a chief series of G we obtain a composition series of G whose successive quotients all have prime order, and hence G is solvable by Proposition 4. Conversely, suppose that G is solvable, and let H/K be a chief factor of G, where $K \lhd H \trianglelefteq G$. Since H/K is solvable by parts (i) and (ii) of Proposition 3, we see from Proposition 4 that every composition factor of H/K is of prime order. Hence H/K is finite; by Corollary 10.6, H/K is isomorphic with a direct product of copies of some simple group S. But then every composition factor of H/K must be isomorphic with S, and as these factors have prime order, we see that H/K is elementary abelian. ∎

COROLLARY 7. A group having a composition series is solvable iff it has a normal series in which every successive quotient is a p-group.

PROOF. Let G be a group having a composition series. Suppose that G has a normal series whose successive quotients are p-groups. We can refine this normal series to a chief series of G, and by doing so we see that each chief factor of G is a section of, and hence is itself, a p-group. But by Corollary 10.6, each chief factor is a direct product of simple groups, and the p-groups that are direct products of simple groups are exactly the elementary abelian p-groups. It now follows from Proposition 6 that G is solvable. The converse follows directly from Proposition 6. ∎

Those finite groups whose chief factors all have prime order are said to be *supersolvable*; finite supersolvable groups are solvable by Proposition 6. It follows from Exercise 8.2 that finite p-groups are supersolvable. However, not all finite solvable groups are supersolvable. The series $\Sigma_4 > A_4 > K > 1$, where K is the Klein four-group, is a chief series of Σ_4, and we have $\Sigma_4/A_4 \cong \mathbf{Z_2}$, $A_4/K \cong \mathbf{Z_3}$, and $K \cong \mathbf{Z_2} \times \mathbf{Z_2}$, which by Proposition 6 shows that Σ_4 is solvable but not supersolvable. There is an equivalent definition of supersolvability which extends to infinite groups; see Exercise 6.

Solvable groups possess a number of properties beyond those which hold for arbitrary groups. The following theorem, due to Philip Hall, is a generalization for finite solvable groups of part (i) of Sylow's theorem.

THEOREM 8. Let G be a finite solvable group of order mn, where m and n are coprime. Then G has a subgroup of order m.

(In other words, this theorem asserts that a finite solvable group possesses Hall subgroups of all possible orders.)

PROOF. Let G be as in the statement of the theorem; we assume by induction that the result holds for any group of order less than $|G|$. Let N be a minimal normal subgroup of G. As seen in the proof of Proposition 10.3, N is a chief factor of G, and hence N is an elementary abelian p-group for some prime p by Proposition 6. As m and n are coprime and $p \mid mn$, p must divide exactly one of m and n. If $p \mid m$, then $|G/N| = (m/|N|)n$ is a product of coprime integers; by induction, G/N has a subgroup of order $m/|N|$, and so by the correspondence theorem G has a subgroup of order m. Now suppose that $p \mid n$. Then by the same argument, G/N has a subgroup H/N of order m. Here $|H| = m|N|$ is a product of coprime integers, and if $H < G$ then by induction H, and hence G, has a subgroup of order m. Thus we assume that $H = G$. But now N is a normal subgroup of G such that $|N| = n$ and $|G : N| = m$ are coprime, and so the result follows from the Schur-Zassenhaus theorem. ∎

This theorem does not hold for all finite groups; for instance, we observed in the proof of Theorem 7.5 that the group A_5 of order $60 = 20 \cdot 3$ has no subgroup of order 20.

P. Hall also proved the following generalization for finite solvable groups of parts (ii) and (iii) of Sylow's theorem, which the reader is asked to prove in the exercises:

THEOREM 9. Let G be a finite solvable group of order mn, where m and n are coprime. Then any two subgroups of G of order m are conjugate, and any subgroup of G whose order divides m is contained in a subgroup of order m. ∎

We now state, without proof, a number of well-known theorems giving conditions under which a finite group is solvable. We start with a classical result that generalizes Corollary 5.

BURNSIDE'S THEOREM. If p and q are primes, then any group of order $p^a q^b$ is solvable. ∎

Burnside's theorem is the best possible result, in the sense that a finite group whose order has exactly three prime divisors need not be solvable, with A_5 providing a counterexample. Burnside proved this theorem in 1904 using the methods of character theory, which will be discussed in Chapter 6; we present a character-theoretic proof of Burnside's theorem in the Appendix. Until the 1960s, there was no proof of this theorem that did not involve character theory.

FEIT-THOMPSON THEOREM. All finite groups of odd order are solvable. ∎

This result, also known as the "odd order theorem," had first been conjectured by Burnside in 1911. As a corollary, we see that the only simple groups of odd order are cyclic groups of prime order. Once it had finally been established that all non-abelian finite simple groups are of even order, the movement toward a classification of all finite simple groups gathered steam. Feit and Thompson completed their proof of this theorem during a special "Finite Group Theory Year," held at the University of Chicago in the school year 1960–61, which brought together the leading minds in the field and thus helped lay the groundwork for the emergence of finite group theory as a highly active area of mathematical research in the 1960s. Their original proof [12] was 255 pages long, and even at that length the proof requires too much background knowledge for it to be readily comprehensible to anyone but the specialists at the time at which the proof was published. In recent years there has been a program, of which [8] represents a major part, to produce a more accessible proof of this fundamental result.

The following result is a converse to Theorem 8 and is also due to P. Hall. While this theorem generalizes Burnside's theorem, its proof relies on the fact that groups of order $p^a q^b$ are solvable.

THEOREM 10. Let G be a finite group. If G has a subgroup of order m whenever m and n are coprime numbers such that $|G| = mn$, then G is solvable. ■

This next result was conjectured by P. Hall in the same paper [14] in which he proved Theorem 10; it was later proved by Thompson in [27].

THEOREM 11. A finite group G is not solvable iff there exist non-trivial elements x, y, z of G of pairwise coprime orders a, b, c such that $xy = z$. ■

For example, in A_5 this criterion for non-solvability is satisfied by taking $a = 5$, $b = 2$, $c = 3$ and $x = (1\ 2\ 3\ 4\ 5)$, $y = (1\ 2)(3\ 4)$, $z = (1\ 3\ 5)$. (Compare with Exercise 1.5.)

The concept of solvable groups originated with Galois around 1830 in his work on the solution by radicals of polynomial equations; this in fact is the origin of the term "solvable" for this class of groups. We now present a theorem of Galois on solvable groups after first introducing the requisite terminology.

Let V be a vector space over a field F. For $v \in V$, let T_v be the self-map of V corresponding to translation by v, and let $\mathcal{T}(V)$ be the set of all such translations; this is a group isomorphic with the abelian group V, and if we regard $\mathcal{T}(V)$ and $\mathrm{GL}(V)$ as subgroups of the group of all invertible self-maps of V, then $\mathrm{GL}(V) \cap \mathcal{T}(V)$ is trivial. We easily verify that $S T_v S^{-1} = T_{S(v)}$ for any $S \in \mathrm{GL}(V)$ and $v \in V$.

This gives rise to a homomorphism from $\mathrm{GL}(V)$ to $\mathrm{Aut}(\mathcal{T}(V))$, and hence we have an (internal) semidirect product $\mathcal{T}(V) \rtimes \mathrm{GL}(V)$. This group is called the *affine group* of V and is denoted $\mathrm{Aff}(V)$; its elements are called the *affine transformations* of V.

THEOREM 12. Let G be a finite, solvable, primitive permutation group on a set X. Then X can be given the structure of a vector space over the field $\mathbb{Z}/p\mathbb{Z}$ for some prime p in such a way so that G is isomorphic with a subgroup of $\mathrm{Aff}(X)$ that contains $\mathcal{T}(X)$.

PROOF. Let N be a minimal normal subgroup of G. Then N is a chief factor of the solvable group G, and hence N is an elementary abelian p-group for some prime p by Proposition 6. We endow N with the structure of a vector space over $\mathbb{Z}/p\mathbb{Z}$ as in the proof of Proposition 4.1. (Recall that for $y, z \in N$, we define $y + z$ to be the product yz in N, which gives a commutative operation since N is abelian, and for $y \in N$ and $\alpha = a + p\mathbb{Z} \in \mathbb{Z}/p\mathbb{Z}$, we define $\alpha y = y^a$, which is well-defined since $y^p = 1$.)

Let $x_0 \in X$ and let H be the stabilizer of x_0. Since X is a primitive G-set, we see from Corollary 3.10 that H is a maximal subgroup of G. Suppose that H contains N. Then N fixes x_0, and $N = gNg^{-1} \leqslant gHg^{-1}$ stabilizes $gx_0 \in X$ for every $g \in G$ by Lemma 3.2. As X is transitive, N now stabilizes every $x \in X$; but the action of G on X is faithful by hypothesis, so we must have $N = 1$, which is a contradiction. Therefore $H < NH \leqslant G$, which forces $NH = G$ since H is maximal. Now $N \cap H \trianglelefteq H$ by Proposition 1.7, and $N \cap H \trianglelefteq N$ since N is abelian; therefore we have $N \cap H \trianglelefteq NH = G$. As N is a minimal normal subgroup of G and $N \cap H \neq N$, we must have $N \cap H = 1$. Therefore $G = N \rtimes H$.

By Proposition 3.4, X is isomorphic as a G-set with G/H. But $G = NH$, so for any $g \in G$ there is some $y \in N$ such that $gH = yH$, and from this it follows that the action of N on X is transitive. Moreover, if $y, z \in N$ are such that $yx_0 = zx_0$, then $z^{-1}y$ lies in $N \cap H = 1$, and hence $y = z$. Thus each element of X can be written as yx_0 for exactly one $y \in N$.

We now give X the structure of a vector space over $\mathbb{Z}/p\mathbb{Z}$ by transferring the structure that we imposed on N. That is, we define $x + x' = (y + y')x_0$ for $x, x' \in X$, where $y, y' \in N$ are the unique elements such that $x = yx_0$, $x' = y'x_0$. Similarly, for $\alpha \in \mathbb{Z}/p\mathbb{Z}$ and $x = yx_0 \in X$ we define $\alpha x = (\alpha y)x_0$. Now if $x = yx_0 \in X$, then $gx = g(yx_0) = (gy)x_0 = (g + y)x_0 = gx_0 + yx_0 = gx_0 + x$ for any $g \in N$, and hence g acts on the vector space X as translation by the vector gx_0. Since X is a transitive N-set, this shows that $N = \mathcal{T}(X)$. It now suffices to show that each element of H acts on X as a linear transformation, a task we leave to the reader. ∎

A normal series $G = G_0 \geqslant G_1 \geqslant \ldots \geqslant G_r = 1$ of a group G is said to be a *central series* of G if, for each i, G_i/G_{i+1} is contained in the center of G/G_{i+1}. A group G is said to be *nilpotent* if it has a central series. An abelian group G has the central series $G > 1$, and hence abelian groups are nilpotent.

PROPOSITION 13. Nilpotent groups are solvable.

PROOF. If G is nilpotent, then it has a central series, which is a normal series with abelian successive quotients, and hence G is solvable by Proposition 2. ∎

There are solvable groups that are not nilpotent. For example, the group Σ_3 cannot have a central series, as the penultimate term of such a series would have to be a non-trivial subgroup of $Z(\Sigma_3) = 1$.

LEMMA 14. Finite p-groups are nilpotent.

PROOF. Let P be a finite p-group. We use induction on $|P|$; if $|P| = p$, then P is abelian and hence nilpotent. Let $Z = Z(P)$. Since $Z \neq 1$ by Theorem 8.1, by induction P/Z has a central series $P/Z = P_0/Z \geqslant P_1/Z \geqslant \ldots \geqslant P_r/Z$, and we see easily that the series $P = P_0 \geqslant P_1 \geqslant \ldots \geqslant P_r = Z \geqslant 1$ is a central series of P. ∎

If H and K are subgroups of a group G, we define a new subgroup $[H, K]$ of G by $[H, K] = <\{[h, k] \mid h \in H, k \in K\}>$. Observe that $G' = [G, G]$.

We now reconcile the above definition of nilpotence with that made in Section 8 for finite groups. Recall that a subgroup H of a group G is said to be self-normalizing if $N_G(H) = H$.

THEOREM 15. The following statements about a finite group G are equivalent:

(1) G is nilpotent.
(2) G has no proper self-normalizing subgroups.
(3) Every Sylow subgroup of G is normal in G.

(4) G is the direct product of its Sylow subgroups.

(5) Every maximal subgroup of G is normal in G.

PROOF. We have $(3) \Leftrightarrow (4) \Leftrightarrow (5)$ by Theorem 8.7, so it suffices to show that $(1) \Rightarrow (2)$, $(2) \Rightarrow (5)$, and $(4) \Rightarrow (1)$.

Suppose that $G = G_0 \geqslant G_1 \geqslant \ldots \geqslant G_r = 1$ is a central series of the nilpotent group G. Let $H < G$, and let k be such that $G_{k+1} \leqslant H$ and $G_k \not\leqslant H$; such a k exists since $G_r = 1$. Clearly $[G_k, H] \leqslant [G_k, G]$. Let $x \in G_k$ and $y \in G$. Since $G_k/G_{k+1} \leqslant Z(G/G_{k+1})$, we find that $[x, y] \in G_{k+1}$; consequently $[G_k, G] \leqslant G_{k+1}$, and so we have $[G_k, H] \leqslant H$. We now see that $G_k \leqslant N_G(H)$; since $G_k \not\leqslant H$, we must have $H < N_G(H)$. Therefore $(1) \Rightarrow (2)$.

Suppose that (2) holds, and let H be a maximal subgroup of G. Since $H < N_G(H)$ by hypothesis, we must have $N_G(H) = G$, and hence $H \triangleleft G$. Therefore $(2) \Rightarrow (5)$.

To show that $(4) \Rightarrow (1)$, it suffices by virtue of Lemma 14 to show that the direct product of two nilpotent groups is nilpotent. We leave this to the reader. ∎

EXERCISES

1. Show that any finite group has a largest solvable normal subgroup.
2. Let N be a solvable normal Hall subgroup of a finite group G. Show that any two complements to N in G are conjugate. (The strong version of the Schur-Zassenhaus theorem asserts that this is true for any normal Hall subgroup N, whether solvable or not. The proof uses the Feit-Thompson theorem to argue that if N is not solvable, then G/N must be solvable.)
3. Complete the following sketch, which gives a proof of Theorem 8 that is independent of the Schur-Zassenhaus theorem. Here G is a finite solvable group of order mn, where m and n are coprime, and we are to show that G has a subgroup of order m.
    (a) Using induction, reduce to the case where G has a unique minimal normal subgroup N of order n.
    (b) Let M/N be a minimal normal subgroup of G/N; since G/N is solvable by Proposition 3, it follows from Proposition 6 that $|M/N|$ is a power of some prime divisor p of m. If P is a Sylow p-subgroup of M, show that $|N_G(P)| = m$.
4. Prove Theorem 9. If your proof involves the Schur-Zassenhaus theorem, then construct a second proof that is independent of Schur-Zassenhaus by mimicking Exercise 3 above.

5. Let V be a finite-dimensional vector space over the field of p elements. Suppose that G is a solvable subgroup of $\mathrm{Aff}(V)$ which contains $\mathcal{T}(V)$ and that the stabilizer of the origin under the action of G on V does not leave any non-zero proper subspace of V invariant. Show that G is a primitive permutation group on V.

6. Show that a finite group is supersolvable iff it has a normal series having cyclic successive quotients. (An arbitrary group is called supersolvable if this latter condition holds.)

7. Show that finite nilpotent groups are supersolvable. (If we define supersolvability for arbitrary groups as in Exercise 6, is an infinite nilpotent group necessarily supersolvable?)

8. (cont.) Give an example of a finite supersolvable group that is not nilpotent. (Hence we have a proper ascending chain of classes of finite groups, starting with cyclic groups, then abelian, then nilpotent, then supersolvable, and ending with solvable.)

FURTHER EXERCISES

Let G be a group. Let $\Gamma_1 = G$, and for $n \in \mathbb{N}$ let $\Gamma_{n+1} = [\Gamma_n, G] \leqslant G$. Let $\mathcal{Z}_0 = 1$, and for $n \in \mathbb{N}$ let $\mathcal{Z}_n$ be the unique subgroup of G such that $\mathcal{Z}_n/\mathcal{Z}_{n-1} = Z(G/\mathcal{Z}_{n-1})$. (Observe that $\Gamma_2 = G'$ and $\mathcal{Z}_1 = Z(G)$.) We call the series $G = \Gamma_1 \geqslant \Gamma_2 \geqslant \Gamma_3 \geqslant \ldots$ the *lower central series* of G and the series $1 = \mathcal{Z}_0 \leqslant \mathcal{Z}_1 \leqslant \mathcal{Z}_2 \leqslant \ldots$ the *upper central series* of G.

9. Show that each Γ_n and $\mathcal{Z}_n$ is a characteristic subgroup of G.

10. (cont.) Show that G is nilpotent iff $\Gamma_r = 1$ for some r iff $\mathcal{Z}_s = G$ for some s.

11. (cont.) Suppose that G is nilpotent. Show that the lower and upper central series of G have the same length, say c, and that no central series of G can have length less than c. (This number c is called the *nilpotency class* of G. The groups of nilpotency class 1 are exactly the abelian groups.)

5
Semisimple Algebras

This chapter provides the algebraic background necessary for the development in Chapter 6 of the character theory of finite groups. Section 12 contains a review of elementary module theory and a treatment of the basic notions of the representation theory of finite groups, finishing with Maschke's theorem. This motivates Section 13, which concentrates on Wedderburn's classification of semisimple algebras and related topics.

12. Modules and Representations

In Section 3 we studied the action of groups on sets, and throughout the book we have seen how useful this notion can be. If a group acts on a set that possesses some additional algebraic structure, then the group action need not behave well with respect to that structure; however, the class of actions that do respect the underlying structure may be of some interest. What we will now focus our attention on are actions of groups on vector spaces that respect the vector space structure.

Let F be a field, and let G be a group acting on an F-vector space V. We say that the action of G on V is *linear* if:

- $g(v + w) = gv + gw$ for all $g \in G$ and $v, w \in V$.
- $g(\alpha v) = \alpha(gv)$ for all $g \in G$, $\alpha \in F$, and $v \in V$.

We found in Proposition 3.1 that group actions correspond to homomorphisms to symmetric groups. We now deduce the analogous
correspondence for linear group actions.

PROPOSITION 1. There is a bijective correspondence between the
set of linear actions of a group G on an F-vector space V and the
set of homomorphisms from G to $\mathrm{GL}(V)$.

PROOF. Suppose that $\rho\colon G \to \mathrm{GL}(V)$ is a homomorphism; then
it is clear that the action of G on V defined by setting $gv = \rho(g)(v)$
is linear. Conversely, if we have a linear action of G on V, then
we can define a homomorphism $\rho\colon G \to \mathrm{GL}(V)$ by $\rho(g)(v) = gv$.
These processes are evidently mutually inverse, establishing the desired correspondence. ∎

A homomorphism $\rho\colon G \to \mathrm{GL}(V)$, where G is a group and V a
vector space, is called a *linear representation* of G in V. We see from
Proposition 1 that the study of linear representations of groups is
equivalent to the study of linear actions of groups. This area of study,
with emphasis on finite groups and finite-dimensional vector spaces,
originated in the late nineteenth century and has proven to have a
number of applications to finite group theory, as well as significant
intrinsic interest.

The modern approach to the representation theory of finite groups
involves yet another equivalent concept, that of finitely generated
modules over group algebras. It is therefore necessary at this junc-
ture to review some elementary module theory. As in Section 1, we
will omit most proofs on the assumption that the reader will have
seen this material previously; however, our account requires only a
mild knowledge of rings, fields, and vector spaces.

Let R be a ring with unit, meaning that R has a multiplicative
identity 1, and let M be an abelian group written additively. We
say that M is a *left R-module* if there is a map from $R \times M$ to M,
with the image of $(r, m) \in R \times M$ being written rm, which satisfies
the following properties:

- $1m = m$ for all $m \in M$.
- $r(m + n) = rm + rn$ for all $r \in R$ and $m, n \in M$.
- $(r + s)m = rm + sm$ for all $r, s \in R$ and $m \in M$.
- $r(sm) = (rs)m$ for all $r, s \in R$ and $m \in M$.

If $R = \mathbb{Z}$, then these conditions are automatically satisfied; that is, $\mathbb{Z}$-modules are exactly the same as abelian groups. If F is a field, then the definition of an F-module is precisely that of an F-vector space. A module, then, is the natural generalization of a vector space when working over an arbitrary ring instead of a field. We can similarly define the notion of a right R-module as a map from $M \times R$ to R, sending (m, r) to mr, which satisfies properties analogous to those above. If R is commutative, then every left R-module can, in an obvious way, be given a right R-module structure, and hence it is not necessary to distinguish between left and right R-modules. (In general, a left R-module can be considered as a right module over the *opposite ring* R^{op}, which is just the abelian group R with the multiplication rule of R reversed; if R is commutative, then R and R^{op} are isomorphic rings.) We will always use R-module to mean left R-module unless otherwise indicated.

Let S be another ring with unit, and suppose that an abelian group M is both a left R-module and a right S-module. We say that M is an (R, S)-*bimodule* if in addition we have $r(ms) = (rm)s$ for every $r \in R$, $m \in M$, and $s \in S$. Any left R-module is an $(R, \mathbb{Z})$-bimodule, and any right R-module is a $(\mathbb{Z}, R)$-bimodule; the ring R is itself an (R, R)-bimodule. If R is commutative, then any R-module is an (R, R)-bimodule.

An R-module M is said to be *finitely generated* if every element of M can be written as an R-linear combination of elements of some finite subset of M. However, minimal generating sets for a given module may have different numbers of elements, which is a stark contrast with the well-known fact that any two finite bases of a given vector space must have the same number of elements.

Let M be an R-module, and let N be a subgroup of M. We say that N is an R-*submodule* (or just *submodule*) of M if $rn \in N$ for every $r \in R$ and $n \in N$. For example, the (left) R-submodules of R are exactly the left ideals of R. Every module has at least two submodules, namely itself and the zero submodule $\{0\}$, which we denote by 0; a module having no other submodules is called *simple*. (As with groups, by convention the zero module is not considered to be simple.) If N is a submodule of M, then since M is an abelian group we may construct the quotient group M/N, and we can give M/N an R-module structure by defining $r(m + N) = rm + N$ for $r \in R$ and $m + N \in M/N$. We call M/N the *quotient R-module* (or just *quotient module*) of M by N.

Let N_1 and N_2 be submodules of an R-module M. We define their *sum* to be $N_1 + N_2 = \{x + y \mid x \in N_1, y \in N_2\} \subseteq M$; this is a submodule of M, as is $N_1 \cap N_2$. If $N_1 \cap N_2 = 0$, then we say that the sum of N_1 and N_2 is *direct*, and we write $N_1 \oplus N_2$ instead of $N_1 + N_2$. We also have an external notion of direct sum: If M and N are R-modules, then we give $M \times N$ an R-module structure via $r(m, n) = (rm, rn)$, and we write $M \oplus N$ instead of $M \times N$. The notions of internal and external direct sums can be extended to any finite number of submodules, as was done in Section 2 for the direct product of groups. We say that a submodule N of a module M is a *direct summand* of M if there is some other submodule N' of M such that $M = N \oplus N'$. In general we use nM to denote the direct sum of n copies of a module M, although we may also write this module as M^n.

A *composition series* of an R-module M is a descending series of submodules of M which terminates in the zero submodule and in which each successive quotient is a simple module. A module need not have a composition series; we observed on page 98 that any $\mathbb{Z}$-module having infinitely many elements does not have a composition series. The analogue of the Jordan-Hölder theorem holds for modules that have composition series, and hence we can speak in a well-defined way about the composition factors of a module. Every composition factor of any submodule or quotient module of an R-module M must also be a composition factor of M, as any composition series of a submodule (resp., quotient module) can be extended (resp., lifted and then extended) to give a composition series of M.

Let M and N be R-modules, and let $\varphi \colon M \to N$ be a group homomorphism. We say that φ is an *R-module homomorphism* if $\varphi(rm) = r\varphi(m)$ for any $r \in R$ and $m \in M$. As always, we define mono-, epi-, iso-, endo-, and automorphisms of R-modules as we did in Section 1 for groups. The *kernel* of φ is the set of elements of M that are mapped under φ to the additive identity of N; it is denoted $\ker \varphi$, and it is a submodule of M. The image of φ is a submodule of N. The fundamental theorem on homomorphisms for modules states that the R-modules $M/\ker \varphi$ and $\operatorname{im} \varphi$ are isomorphic via the map induced by φ; again, this is exactly analogous to the group-theoretic case. There are also module-theoretic analogues of the correspondence theorem and the two main isomorphism theorems. For example, the first isomorphism theorem for modules states that if M is an R-module having submodules N_1 and N_2, then $N_1 + N_2/N_1 \cong N_2/N_1 \cap N_2$.

SCHUR'S LEMMA. Any non-zero homomorphism between simple R-modules is an isomorphism.

PROOF. Let M and N be simple R-modules, and let $\varphi \colon M \to N$ be an R-module homomorphism. As $\ker \varphi$ is a submodule of M, we must have either $\ker \varphi = M$, in which case $\varphi = 0$, or $\ker \varphi = 0$; similarly, $\operatorname{im} \varphi$ is a submodule of N, so either $\operatorname{im} \varphi = 0$, forcing $\varphi = 0$, or $\operatorname{im} \varphi = N$. Hence if $\varphi \neq 0$, then φ is an isomorphism. ∎

If M and N are R-modules, then we denote the set of all R-module homomorphisms from M to N by $\operatorname{Hom}_R(M, N)$. We write $\operatorname{End}_R(M)$ for $\operatorname{Hom}_R(M, M)$. We give $\operatorname{Hom}_R(M, N)$ an abelian group structure as follows: For $\varphi, \rho \in \operatorname{Hom}_R(M, N)$, we define $\varphi + \rho \in \operatorname{Hom}_R(M, N)$ by $(\varphi + \rho)(m) = \varphi(m) + \rho(m)$ for all $m \in M$.

Let S be another ring with unit, let M be an (R, S)-bimodule, and let N be an R-module. For $s \in S$ and $\varphi \in \operatorname{Hom}_R(M, N)$, we define $s\varphi \in \operatorname{Hom}_R(M, N)$ by $(s\varphi)(m) = \varphi(ms)$. With this definition, $\operatorname{Hom}_R(M, N)$ becomes an S-module. For example, if F is a field and U and V are F-vector spaces, then $\operatorname{Hom}_F(U, V)$ is also an F-vector space via this definition; in this case, $\lambda \in F$ acts on $\varphi \in \operatorname{Hom}_F(U, V)$ by $(\lambda\varphi)(u) = \varphi(\lambda u)$ for $u \in U$. Note that the map from $\operatorname{Hom}_R(R, M)$ to M sending φ to $\varphi(1)$ is an isomorphism of R-modules.

Let M be an (R, S)-bimodule and let N be an S-module. We say that a set map f from $M \times N$ to an R-module U is *balanced* if:

- $f(m_1 + m_2, n) = f(m_1, n) + f(m_2, n)$ for all $m_1, m_2 \in M$ and $n \in N$.
- $f(m, n_1 + n_2) = f(m, n_1) + f(m, n_2)$ for all $m \in M$ and $n_1, n_2 \in N$.
- $f(ms, n) = f(m, sn)$ for all $m \in M$, $n \in N$, and $s \in S$.
- $f(rm, n) = rf(m, n)$ for all $m \in M$, $n \in N$, and $r \in R$.

The *tensor product* of M and N over S is an R-module, denoted $M \otimes_S N$, equipped with a balanced map $\eta \colon M \times N \to M \otimes_S N$ with the property that if U is an R-module and $f \colon M \times N \to U$ is a balanced map, then there is a unique R-module homomorphism $\alpha \colon M \otimes_S N \to U$ such that $f = \alpha \circ \eta$. Tensor products exist and are unique up to isomorphism. We write $m \otimes n = \eta(m, n)$ for any $m \in M$ and $n \in N$.

More concretely, the tensor product $M \otimes_S N$ is the R-module generated by the set $\{m \otimes n \mid m \in M, \, n \in N\}$, where the symbols $m \otimes n$ satisfy the following identities:

- $(m_1 + m_2) \otimes n = m_1 \otimes n + m_2 \otimes n$ for all $m_1, m_2 \in M$ and $n \in N$.
- $m \otimes (n_1 + n_2) = m \otimes n_1 + m \otimes n_2$ for all $m \in M$ and $n_1, n_2 \in N$.
- $(ms) \otimes n = m \otimes (sn)$ for all $m \in M$, $n \in N$, and $s \in S$.
- $(rm) \otimes n = r(m \otimes n)$ for all $m \in M$, $n \in N$, and $r \in R$.

Observe that an arbitrary element of $M \otimes_S N$ is not a symbol $m \otimes n$, but rather a sum of such symbols.

For example, consider the special case where F is a field and U and V are finite-dimensional F-vector spaces. Then U is an (F, F)-bimodule, so we can construct the F-vector space $U \otimes_F V$. If $\{u_1, \ldots, u_r\}$ and $\{v_1, \ldots, v_s\}$ are bases for U and V, respectively, then $U \otimes_F V$ is an rs-dimensional F-vector space having as a basis the set $\{u_i \otimes v_j \mid 1 \leq i \leq r, \, 1 \leq j \leq s\}$. If $u = \sum_i a_i u_i \in U$ and $v = \sum_j b_j v_j \in V$, then we have $u \otimes v = \sum_{i,j} a_i b_j (u_i \otimes v_j)$. (An understanding of this special case will, strictly speaking, suffice in reading the remainder of the book; however, an understanding of the general case will be beneficial in Section 16.)

PROPOSITION 2. Let R be a ring with unit, and let M be an R-module. Then M and $R \otimes_R M$ are isomorphic R-modules.

PROOF. The map $f \colon R \times M \to M$ defined by $f(r, m) = rm$ is easily seen to be balanced and hence induces an R-module homomorphism $\alpha \colon R \otimes_R M \to M$ where $\alpha(r \otimes m) = rm$. The map α has as its inverse the R-module homomorphism sending $m \in M$ to $1 \otimes m$. ∎

PROPOSITION 3. Let R and S be rings with unit, let $M_1, \ldots, M_r$ be (R, S)-bimodules, and let N be an S-module. Then $(\oplus_i M_i) \otimes_S N$ and $\oplus_i M_i \otimes_S N$ are isomorphic R-modules.

PROOF. We define a map $f \colon (\oplus_i M_i) \times N \to \oplus_i M_i \otimes_S N$ by $f((m_1, \ldots, m_r), n) = (m_1 \otimes n, \ldots, m_r \otimes n)$. We see that f is balanced and hence induces a homomorphism from $(\oplus_i M_i) \otimes_S N$ to $\oplus_i M_i \otimes_S N$. By similar reasoning, we can define a map that will be inverse to the above homomorphism; we leave the details to the reader. ∎

PROPOSITION 4. Let F be a field, and let U and V be F-vector spaces, with $\dim_F(U) < \infty$. Let $U^* = \operatorname{Hom}_F(U, F)$. Then the map $\Gamma: U^* \otimes_F V \to \operatorname{Hom}_F(U, V)$ defined by setting $\Gamma(\varphi \otimes v)(u) = \varphi(u)v$ and extending linearly is an isomorphism of F-vector spaces.

PROOF. We observe that Γ is the homomorphism induced by the balanced map sending $(\varphi, v) \in U^* \times V$ to the linear transformation that maps $u \in U$ to $\varphi(u)v \in V$. Let $\{u_1, \ldots, u_n\}$ be a basis for U. For each $1 \leq i \leq n$, let $f_i \in U^*$ be defined by $f_i(u_j) = \delta_{ij}$ for each j. Then $\{f_1, \ldots, f_n\}$ is a basis for U^*. We can write an arbitrary element $\sum_k (\sum_i \alpha_{ik} f_i) \otimes v_k$ of $U^* \otimes F$ as $\sum_i f_i \otimes (\sum_k \alpha_{ik} v_k)$; hence each element of $U^* \otimes F$ has the form $\sum_{i=1}^n f_i \otimes v_i$ for some $v_i \in V$.

Let $\sum_{i=1}^n f_i \otimes v_i \in U^* \otimes_F V$, and fix some j. Then we have

$$\Gamma(\sum_{i=1}^n f_i \otimes v_i)(u_j) = \sum_{i=1}^n \Gamma(f_i \otimes v_i)(u_j) = \sum_{i=1}^n f_i(u_j)v_i = v_j.$$

Therefore, if $\Gamma(\sum_{i=1}^n f_i \otimes v_i) = 0$, then $\sum_{i=1}^n f_i \otimes v_i = 0$ since $v_j = 0$ for all j, showing that Γ is injective. Now let $\sigma \in \operatorname{Hom}_F(U, V)$. Then for $u = \sum_{i=1}^n \alpha_i u_i \in U$, we have

$$\Gamma(\sum_{i=1}^n f_i \otimes \sigma(u_i))(u) = \sum_{i=1}^n f_i(u)\sigma(u_i) = \sum_{i=1}^n \alpha_i \sigma(u_i) = \sigma(u),$$

which shows that $\Gamma(\sum_{i=1}^n f_i \otimes \sigma(u_i)) = \sigma$ and hence that Γ is surjective. ∎

Having reviewed enough module theory for our purposes, we now need to introduce the class of rings whose modules we will be studying. Let R be a ring and let G be a group. The *group ring* of G over R, denoted by RG, consists of all finite formal R-linear combinations of elements of G, with the obvious rule for addition and with multiplication defined by extending the multiplication in G; explicitly, we have

$$(\sum_{x \in G} \alpha_x x)(\sum_{y \in G} \beta_y y) = \sum_{x \in G}\sum_{y \in G} \alpha_x \beta_y xy = \sum_{x \in G}(\sum_{g \in G} \alpha_g \beta_{g^{-1}x})x.$$

The group ring has a unit element, namely the identity element of the underlying group. We will be primarily interested in the case where $R = F$ is a field and G is finite, in which case FG is not only a ring but also an F-vector space having G as a basis and hence having finite dimension $|G|$. In this case, FG is called the *group algebra* rather than the group ring, since it is an example of an algebra, a mathematical structure we now define. (Algebras will be the main objects of study in Section 13.)

If F is a field, then an *algebra* over F (or simply an F-algebra) is a set A with a ring structure and an F-vector space structure that share the same addition operation, and with the additional property that $(\lambda a)b = \lambda(ab) = a(\lambda b)$ for any $\lambda \in F$ and $a, b \in A$. (We do not assume that an algebra is necessarily a ring with unit.) An algebra is called *finite-dimensional* if it has finite dimension as an F-vector space. For example, the matrix ring $\mathcal{M}_n(F)$ is a finite-dimensional F-algebra for any $n \in \mathbb{N}$, and we have seen that FG is a finite-dimensional F-algebra when G is finite. A *homomorphism of F-algebras* is a ring homomorphism which is also an F-linear transformation.

Modules over a group algebra FG can also be regarded as F-vector spaces, with $\lambda \in F$ acting by $\lambda 1 \in FG$, and when G is finite we have the following nice relationship between these structures:

LEMMA 5. If F is a field and G a finite group, then an FG-module is finitely generated iff it has finite dimension as an F-vector space.

PROOF. If V is generated as an FG-module by $\{v_1, \ldots, v_t\}$, then V is a generated as an F-vector space by $\{gv_i \mid g \in G, 1 \le i \le t\}$, and since G is finite we see that $\dim_F(V) < \infty$. The converse is trivial. ∎

We now derive the fundamental connection between modules over group algebras and representation theory.

PROPOSITION 6. If F is a field and G a finite group, then there is a bijective correspondence between finitely generated FG-modules and linear actions of G on finite-dimensional F-vector spaces.

PROOF. If V is a finitely generated FG-module, then $\dim_F(V)$ is finite by Lemma 5, and the map from $G \times V$ to V obtained by restricting the module structure map from $FG \times V$ to V is clearly a linear action. Conversely, suppose that V is a finite-dimensional F-vector space on which G acts linearly; then we give V an FG-module structure by defining $(\sum_{g \in G} \alpha_g g)v = \sum_{g \in G} \alpha_g(gv)$ for $\sum_{g \in G} \alpha_g g \in FG$ and $v \in V$. It is clear that these processes are mutually inverse. ∎

We see from the above proof that, in order to define an FG-module structure on an F-vector space V, it suffices to stipulate the action of elements of G on V; the action of arbitrary elements of FG on V is then defined by a linear extension process. We shall do this implicitly when defining modules over group algebras.

We now embark on our study of the representation theory of finite groups. In the remainder of this section, G will denote a finite group and F a field; all F-vector spaces will be finite-dimensional, and all FG-modules will be finitely generated (and hence of finite F-dimension by Lemma 5). Our viewpoint will primarily be that of modules over the group algebra, although on occasion it will be notationally beneficial to work with the linear representation ρ arising from a given FG-module V, where $\rho\colon G \to \mathrm{GL}(V)$ is defined by $\rho(g)(v) = gv$ for $g \in G$ and $v \in V$.

We start by considering some elementary examples. The field F can always be regarded as an FG-module by defining $g\lambda = \lambda$ for all $g \in G$ and $\lambda \in F$. This module is called the *trivial module*. Now suppose that G acts on a finite set X, and let FX be the set of all formal F-linear combinations of elements of X. This set has an obvious F-vector space structure, with basis X. We define an FG-module structure on FX by linearly extending the action of G on X. For instance, if $X = \{x_1, \ldots, x_n\}$, then for $g \in G$ and $\sum_i c_i x_i \in FX$ we have $g(\sum_i c_i x_i) = \sum_i c_i(gx_i)$. Such modules are called *permutation modules*.

If U and V are FG-modules, then we already know that $U \oplus V$ has a natural FG-module structure, given by $g(u,v) = (gu, gv)$; and we also know that we can construct the F-vector spaces $U \otimes_F V$ and $\mathrm{Hom}_F(U,V)$. What we now show is that these latter vector spaces also admit natural FG-module structures. Each element $g \in G$ defines a balanced map from $U \times V$ to $U \otimes_F V$ sending (u,v) to $gu \otimes gv$, and from this we find that $U \otimes_F V$ becomes an FG-module under the so-called "diagonal action" given by $g(u \otimes v) = gu \otimes gv$ and linear extension. Now for $g \in G$ and $\varphi \in \mathrm{Hom}_F(U,V)$, we define $g\varphi\colon U \to V$ by $(g\varphi)(u) = g(\varphi(g^{-1}u))$. It is easily verified that $g\varphi \in \mathrm{Hom}_F(U,V)$, and for $g_1, g_2 \in G$ we have

$$((g_1 g_2)\varphi)(u) = g_1 g_2 \varphi((g_1 g_2)^{-1}u) = g_1(g_2\varphi(g_2^{-1}(g_1^{-1}u)))$$

$$= g_1((g_2\varphi)(g_1^{-1}u)) = (g_1(g_2\varphi))(u),$$

which shows that $(g_1 g_2)\varphi = g_1(g_2\varphi)$. This gives $\mathrm{Hom}_F(U,V)$ an

FG-module structure. We write U^* for $\mathrm{Hom}_F(U, F)$, where F is the trivial module, and we call U^* the *dual module* of U; here we have $(g\varphi)(u) = \varphi(g^{-1}u)$.

PROPOSITION 7. Let U and V be FG-modules. Then $U^* \otimes_F V$ and $\mathrm{Hom}_F(U, V)$ are isomorphic FG-modules.

PROOF. By Proposition 4, the map $\Gamma: U^* \otimes_F V \to \mathrm{Hom}_F(U, V)$ defined by $\Gamma(\varphi \otimes v)(u) = \varphi(u)v$ is an isomorphism of F-vector spaces. To show that Γ is an FG-module isomorphism, it suffices to show that $\Gamma(g(\varphi \otimes v))(u)$ and $[g\Gamma(\varphi \otimes v)](u)$ are equal for any $g \in G$, $\varphi \in U^*$, $v \in V$, and $u \in U$; we leave it to the reader to show that they are both equal to $\varphi(g^{-1}u)gv$. ∎

We now establish the most basic result of the representation theory of finite groups. It was discovered in 1898 by Heinrich Maschke, the most junior of the three initial mathematics faculty of the University of Chicago.

MASCHKE'S THEOREM. Let G be a group, and suppose that the characteristic of F is either zero or coprime to $|G|$. If U is an FG-module and V is an FG-submodule of U, then V is a direct summand of U as FG-modules.

(We shall summarize the hypothesis made in Maschke's theorem by saying that the characteristic of F does not divide $|G|$, even though this is a slight abuse of language.)

PROOF. Since V is in particular an F-vector subspace of U, we know from linear algebra that there is some subspace W of U such that $U = V \oplus W$ as F-vector spaces; however, W may not be an FG-submodule of U. Let $\pi: U \to V$ be the projection of U onto V along W, so that π is the unique linear transformation that is the identity on V and zero on W. We define a linear transformation $\pi': U \to U$ by

$$\pi'(u) = \frac{1}{|G|} \sum_{g \in G} g\pi(g^{-1}u)$$

for $u \in U$. (For this definition to make sense, we require that $|G| \neq 0$ in F, which is equivalent to our hypothesis concerning the characteristic of F.)

As V is a submodule of U, we have $gv \in V$ for any $g \in G$ and $v \in V$, and from this we see that π' maps U into V. Also, since π is the identity on V, we see that $g\pi(g^{-1}v) = gg^{-1}v = v$ for any $g \in G$ and $v \in V$; therefore, the restriction of π' to V is the identity. It now follows from linear algebra that $U = V \oplus \ker \pi'$ as F-vector spaces.

It remains to show only that $\ker \pi'$ is an FG-submodule of U. To show this, it suffices to show that π' is an FG-module homomorphism; thus, we must show that $\pi'(xu) = x\pi'(u)$ for any $x \in G$ and $u \in U$. We have

$$\pi'(xu) = \frac{1}{|G|} \sum_{g \in G} g\pi(g^{-1}xu) = \frac{1}{|G|} \sum_{g \in G} xx^{-1}g\pi(g^{-1}xu)$$

$$= \frac{1}{|G|}x\Big(\sum_{g \in G} x^{-1}g\pi(g^{-1}xu)\Big).$$

But as g varies through G, $y = x^{-1}g$ also varies through G for fixed $x \in G$; therefore, after reindexing we have

$$\pi'(xu) = x\Big(\frac{1}{|G|} \sum_{y \in G} y\pi(y^{-1}u)\Big) = x\pi'(u)$$

as required. ∎

A module is said to be *semisimple* if it is a direct sum of simple modules.

COROLLARY 8. Let G be a group, and let F be a field whose characteristic does not divide $|G|$. Then every non-zero FG-module is semisimple.

PROOF. Let U be a non-zero FG-module. We will use induction on $\dim_F(U)$. If U is simple, then we are done; this includes the case $\dim_F(U) = 1$. Hence, we assume that $\dim_F(U) > 1$ and that U is not simple, so that U must have a non-zero proper submodule V. By Maschke's theorem, we have $U = V \oplus W$ for some non-zero proper submodule W of U. But both V and W must have dimension strictly less than that of U, and hence are semisimple by induction; therefore, U is also semisimple. ∎

In Chapter 6 we shall be concerned only with the case $F = \mathbb{C}$, which is called *ordinary* representation theory. Since $\mathbb{C}$ has characteristic zero, we see from Corollary 8 that every non-zero $\mathbb{C}G$-module is semisimple for any group G. The next section concentrates on algebras that have this property. The study of FG-modules when the characteristic of F divides $|G|$, in which case arbitrary FG-modules need not be semisimple, is called *modular* representation theory; its study was originated by Dickson in 1907 (although most of the main developments were made by Richard Brauer starting in the 1930s), and it has a different flavor than what we shall study in Chapter 6.

EXERCISES

Throughout these exercises, let F be a field and let G be a finite group.

1. Let U and V be FG-modules having the same dimension n, and let $\rho: G \to \mathrm{GL}(U)$ and $\tau: G \to \mathrm{GL}(V)$ be the corresponding representations. By fixing F-bases for U and V, we consider ρ and τ as homomorphisms from G to $\mathrm{GL}(n, F)$. Show that $U \cong V$ iff there exists some $M \in \mathrm{GL}(n, F)$ such that $\rho(g)M = M\tau(g)$ for every $g \in G$. (Two representations are said to be *equivalent* if this latter condition holds; hence two representations are equivalent iff their corresponding modules over the group algebra are isomorphic.)

2. Let $\sigma = \sum_{g \in G} g \in FG$. Show that the subspace $F\sigma$ of FG is the unique submodule of FG that is isomorphic with the trivial module.

3. (cont.) Let $\epsilon: FG \to F$ be the FG-module epimorphism defined by $\epsilon(g) = 1$ for all $g \in G$, and let $\Delta = \ker \epsilon$. (We call Δ the *augmentation ideal* of FG.) Show that Δ is the unique submodule of FG whose quotient is isomorphic with the trivial module.

4. (cont.) Suppose that the characteristic of F divides $|G|$. Show that $F\sigma \subseteq \Delta$. Conclude that Δ is not a direct summand of FG and hence that the FG-module FG is not semisimple. (This shows that the converse of Corollary 8 is true.)

FURTHER EXERCISES

In Section 9 we defined the second cohomology group of a pair (H, A), where H is a group and A an abelian group, with respect to a given homomorphism from H to $\mathrm{Aut}(A)$. It is easy to verify that, for an abelian group A, specifying a homomorphism from a group H to $\mathrm{Aut}(A)$ is exactly the same as specifying a $\mathbb{Z}H$-module structure on A. Therefore, it is customary to talk about the second cohomology group of a pair (H, A) where H is a group and A a $\mathbb{Z}H$-module. (We anticipated this development by writing xa instead of $\varphi(x)(a)$ on page 86.) We will now generalize the second cohomology group.

Let H be a group, let H^n be the set of n-tuples of elements of H for some $n \in \mathbb{N}$, and let A be a $\mathbb{Z}H$-module. A *normalized n-cochain* of (H, A) is a set function $f: H^n \to A$ such that $f(h_1, \ldots, h_n) = 0$ whenever $h_i = 1$ for some i; the set of normalized n-cochains of (H, A) is denoted $C^n(H, A)$.

We give $C^n(H, A)$ an abelian group structure by defining

$$(f + g)(h_1, \ldots, h_n) = f(h_1, \ldots, h_n) + g(h_1, \ldots, h_n).$$

We define a homomorphism $\delta^n \colon C^{n-1}(H, A) \to C^n(H, A)$ by

$$(\delta^n f)(h_1, \ldots, h_n) = h_1 f(h_2, \ldots, h_n) + \sum_{i=1}^{n-1} (-1)^i f(h_1, \ldots, h_i h_{i+1}, \ldots, h_n)$$
$$+ (-1)^n f(h_1, \ldots, h_{n-1}).$$

The image of δ^n is called the set of *n-coboundaries* and is denoted $B^n(H, A)$; the kernel of δ^{n+1} is called the set of *n-cocycles* and is denoted $Z^n(H, A)$. For example, ker δ^3 consists of all normalized 2-cochains f satisfying $h_1 f(h_2, h_3) - f(h_1 h_2, h_3) + f(h_1, h_2 h_3) - f(h_1, h_2) = 0$ for all $h_1, h_2, h_3 \in H$, which agrees with our definition of a 2-cocycle in Section 9. Both $B^n(H, A)$ and $Z^n(H, A)$ are abelian subgroups of $C^n(H, A)$. We find that $\delta^{n+1} \circ \delta^n$ is the zero map, or equivalently that $B^n(H, A) \subseteq Z^n(H, A)$. (Verify this.) The quotient group $Z^n(H, A)/B^n(H, A)$ is denoted $H^n(H, A)$ and is called the *nth cohomology group* of (H, A).

While $H^3(H, A)$ plays some role in the theory of extensions of a non-abelian group N by H, where $Z(N) = A$ (see [20, pp. 124–131]), the groups $H^n(H, A)$ for $n > 3$ have no known group-theoretic meaning. Nonetheless, higher cohomology groups are studied for their own sake, and in recent years connections between the cohomology of groups and the representation theory of groups have attracted much attention. (As the title of one of the first author's papers [6] proclaims, "Cohomology *is* representation theory.") We developed the group-theoretic meaning of $H^2(H, A)$ in the further exercises to Section 9, and in what follows we develop the meaning of $H^1(H, A)$.

5. (cont.) Define a homomorphism $\varphi \colon H \to \operatorname{Aut}(A)$ by $\varphi(x)(a) = xa$. Show that there is a bijective correspondence between the set of splitting maps of $A \rtimes_\varphi H$ and $Z^1(H, A)$. (Recall that a homomorphism $t \colon H \to A \rtimes_\varphi H$ is called a splitting map if $\eta \circ t$ is the identity map on H, where $\eta \colon A \rtimes_\varphi H \to H$ is the natural map. The natural inclusion of H into $A \rtimes_\varphi H$ is a splitting map, but there may be others.)

6. (cont.) We say that two splitting maps t and u of $A \rtimes_\varphi H$ are *conjugate* if there is some $a \in A$ such that $u(x) = at(x)a^{-1}$ for all $x \in H$. (Verify that this is an equivalence relation on the set of splitting maps.) Show that there is a bijective correspondence between the set of conjugacy classes of splitting maps of $A \rtimes_\varphi H$ and $H^1(H, A)$.

13. Wedderburn Theory

We saw as a consequence of Maschke's theorem that if F is a field whose characteristic does not divide the order of a finite group G, then every finitely generated FG-module can be written as a direct sum of (finitely many) simple modules. In this section, we investigate the structure of finite-dimensional algebras that have this property; this will both allow us to finish the proof of Kolchin's theorem on unipotent subgroups of general linear groups and provide us with key information about complex group algebras of finite groups, which are the focus of Chapter 6. The results of this section are primarily due to J. H. M. Wedderburn, a Scottish mathematician whose work in the theory of algebras had its genesis when, at the age of 22, he spent the school year 1904–05 as a visiting scholar at the University of Chicago.

All algebras in this section will be finite-dimensional F-algebras, where F is an arbitrary field, and unless explicitly stated otherwise will be algebras with unit. All modules over algebras are assumed to be finitely generated, or equivalently (arguing as in Lemma 12.5) finite-dimensional as F-vector spaces. All direct sums of modules are assumed to be finite.

Let A be an algebra. Our interest is in semisimple A-modules and in determining conditions on A under which every A-module will be semisimple.

LEMMA 1. The following statements about an A-module M are equivalent:

(1) Any submodule of M is a direct summand of M.
(2) M is semisimple.
(3) M is a sum (but not, a priori, a direct sum) of simple submodules.

PROOF. The proof that $(1) \Rightarrow (2)$ is implicit in the proof of Corollary 12.8, and so we shall not repeat it. As the implication $(2) \Rightarrow (3)$ is immediate, it suffices to show that $(3) \Rightarrow (1)$.

Suppose that (3) holds. Let N be a submodule of M, and let V be a submodule of M that is maximal among all submodules of M that intersect N trivially; we wish to show that $N + V = M$. Suppose that $N + V \subset M$. If every simple submodule of M were contained in $N + V$, then as M can be written as a sum of simple submodules, we would have $M \subseteq N + V$. This is not the case, so there is some simple submodule S of M that is not contained in $N + V$. Since $S \cap (N + V)$ is a proper submodule of the simple module S, we must have $S \cap (N + V) = 0$. In particular $S \cap V = 0$, so we have $V \subset V + S$. Let $n \in N \cap (V + S)$; then $n = v + s$ for some $v \in V$ and $s \in S$. This gives $s = n - v \in S \cap (N + V)$, and hence $s = 0$; thus $n = v$, which forces $n = 0$ since $N \cap V = 0$. Therefore $N \cap (V + S) = 0$, which contradicts the maximality of V. We now have $M = N + V$, and since $N \cap V = 0$ we see that M is the direct sum of N and V and hence that N is a direct summand of M. Therefore $(3) \Rightarrow (1)$. ∎

LEMMA 2. Submodules and quotient modules of semisimple modules are semisimple.

PROOF. Let M be a semisimple A-module. By Lemma 1 and the first homomorphism theorem for modules, we see that every submodule of M is isomorphic with a quotient module of M; thus it suffices to show that quotient modules of M are semisimple. Let M/N be an arbitrary quotient module, and let $\eta: M \to M/N$ be the natural map. As M is semisimple, we see from Lemma 1 that M is a sum of simple submodules, say $M = S_1 + \ldots + S_n$. Then we must have $M/N = \eta(M) = \eta(S_1) + \ldots + \eta(S_n)$; but each $\eta(S_i)$ is isomorphic with a quotient module of S_i and thus must be either zero or simple. Therefore, M/N is a sum of simple modules and hence is semisimple by Lemma 1. ∎

We shall say that the algebra A is *semisimple* if all non-zero A-modules are semisimple. If G is a finite group and the characteristic of F does not divide $|G|$, then FG is semisimple by Corollary 12.8. We now present some basic results on semisimple algebras.

LEMMA 3. The algebra A is semisimple iff the A-module A is semisimple.

PROOF. Suppose that the A-module A is semisimple, and let M be an A-module generated by $\{m_1, \ldots, m_r\}$. Let A^r denote the direct sum of r copies of A. We define a map from A^r to M by sending $(a_1, \ldots, a_r)$ to $a_1 m_1 + \ldots + a_r m_r$; this map is an A-module epimorphism. Thus, M is isomorphic with a quotient module of the semisimple module A^r and hence is semisimple by Lemma 2. It follows that A is a semisimple algebra. The converse is trivial. ∎

PROPOSITION 4. Let A be a semisimple algebra, and suppose that as A-modules we have $A \cong S_1 \oplus \ldots \oplus S_r$ where the S_i are simple submodules of A. Then any simple A-module is isomorphic with some S_i.

PROOF. Let S be a simple A-module, fix some $0 \neq s \in S$, and define an A-module homomorphism $\varphi \colon A \to S$ by $\varphi(a) = as$ for $a \in A$. As S is simple, φ is surjective. For each i, let $\varphi_i \colon S_i \to S$ be the restriction of φ to S_i. If $\varphi_i = 0$ for all i, then we would have $\varphi = 0$; hence φ_i is non-zero for some i, and it now follows from Schur's lemma that $\varphi_i \colon S_i \to S$ is an isomorphism. ∎

PROPOSITION 5. Suppose that A is a semisimple algebra, and let $S_1, \ldots, S_r$ be a collection of simple A-modules such that every simple A-module is isomorphic with exactly one S_i. Let M be an A-module, and write $M \cong n_1 S_1 \oplus \ldots \oplus n_r S_r$ for some non-negative integers n_i. Then the n_i are uniquely determined.

(Whenever the modules $S_1, \ldots, S_r$ are as stated in the theorem, we shall say that the S_i are the distinct simple A-modules.)

PROOF. There is a composition series of $n_1 S_1 \oplus \ldots \oplus n_r S_r$ having $n_1 + \ldots + n_r$ terms, in which each S_i appears n_i times as a composition factor; the result now follows from the Jordan-Hölder theorem for modules. ∎

We now initiate our efforts to classify all semisimple algebras. We start by proving the semisimplicity of a certain class of algebras, and we will ultimately show that all semisimple algebras lie in this class.

If D is a finite-dimensional F-algebra, then for any $n \in \mathbb{N}$ the set $\mathcal{M}_n(D)$ of $n \times n$ matrices with entries in D is a finite-dimensional F-algebra of dimension $n^2 \dim_F D$. Algebras of the form $\mathcal{M}_n(D)$ are called matrix algebras over D. For $1 \leq i, j \leq n$ and $\alpha \in D$, let $E_{ij}(\alpha)$ be the matrix whose only non-zero entry occurs in the (i, j)-position and is equal to α. Let D^n be the set of column vectors of length n with entries in D; this forms an $\mathcal{M}_n(D)$-module under matrix multiplication.

An algebra D is said to be a *division algebra* if the non-zero elements of D form a group under multiplication. Any field extension of F is a division algebra, but there may be division algebras that are non-commutative rings.

THEOREM 6. Let D be a division algebra, and let $n \in \mathbb{N}$. Then any simple $\mathcal{M}_n(D)$-module is isomorphic with D^n, and $\mathcal{M}_n(D)$ is isomorphic as $\mathcal{M}_n(D)$-modules with the direct sum of n copies of D^n. In particular, $\mathcal{M}_n(D)$ is a semisimple algebra.

PROOF. A non-zero submodule of D^n must contain some non-zero vector, which must have a non-zero (and hence invertible) entry x in the jth place for some j. By premultiplying this vector by $E_{jj}(x^{-1})$, we see that the submodule contains the jth standard basis vector. By premultiplying this basis vector by appropriate permutation matrices, we see that the submodule contains every standard basis vector, and hence contains every vector. Therefore D^n is the only non-zero $\mathcal{M}_n(D)$-submodule of D^n, and hence D^n is simple. Now for each $1 \leq k \leq n$, let C_k be the submodule of $\mathcal{M}_n(D)$ consisting of those matrices whose only non-zero entries appear in the kth column. Then we clearly have $\mathcal{M}_n(D) \cong \oplus_{k=1}^n C_k$ as $\mathcal{M}_n(D)$-modules; but each C_k is isomorphic as an $\mathcal{M}_n(D)$-module with D^n. It now follows from Lemma 3 that $\mathcal{M}_n(D)$ is a semisimple algebra and from Proposition 4 that D^n is the unique simple $\mathcal{M}_n(D)$-module. ∎

We say that an algebra is *simple* if its only two-sided ideals are itself and the zero ideal.

LEMMA 7. Simple algebras are semisimple.

PROOF. Let A be a simple algebra, and let Σ be the sum of all simple submodules of A. Let S be a simple submodule of A, and let $a \in A$. Then Sa is the image of S under the homomorphism that sends s to sa, and thus Sa is either zero or simple. In either case, we have $Sa \subseteq \Sigma$ for any simple submodule S and any $a \in A$; from this we conclude that Σ is a right ideal in A and hence that Σ is a two-sided ideal. But A is simple and $\Sigma \neq 0$, so we must have $\Sigma = A$. Therefore, A is a sum of simple A-modules, and it follows from Lemmas 1 and 3 that A is a semisimple algebra. ∎

THEOREM 8. Let D be a division algebra, and let $n \in \mathbb{N}$. Then $\mathcal{M}_n(D)$ is a simple algebra.

PROOF. Let $0 \neq M \in \mathcal{M}_n(D)$; we must show that the principal two-sided ideal J of $\mathcal{M}_n(D)$ generated by M is equal to $\mathcal{M}_n(D)$. It suffices to show that J contains each $E_{ij}(1)$ since these matrices generate $\mathcal{M}_n(D)$ as an $\mathcal{M}_n(D)$-module. As $M \neq 0$, there are some $1 \leq r, s \leq n$ such that the (r, s)-entry of M is non-zero; call this entry x. We easily verify that $E_{ss}(1) = E_{sr}(x^{-1})ME_{ss}(1) \in J$. Now let $1 \le i, j \le n$, and let w and w' be the permutation matrices corresponding to the transpositions $(i \ s)$ and $(s \ j)$, respectively. Then $E_{ij}(1) = wE_{ss}(1)w' \in J$. ■

If $B_1, \ldots, B_r$ are algebras, then their *external direct sum* is the algebra B whose underlying set is the Cartesian product of the B_i and whose addition, multiplication, and scalar multiplication operations are defined componentwise. As the name suggests, we write $B = B_1 \oplus \ldots \oplus B_r$. If M is a B_i-module for some i, then we give M a B-module structure by $(b_1, \ldots, b_r)m = b_i m$. Clearly, if M is simple (resp., semisimple) as a B_i-module, then it is also simple (resp., semisimple) as a B-module. For each i, the set of elements of B whose only non-zero entry occurs in the ith component is an ideal of B, and this ideal is isomorphic, as a B-module, with B_i.

Now suppose that B is an algebra having ideals $B_1, \ldots, B_r$ such that, as vector spaces, B equals the direct sum of the B_i. Then B is isomorphic with the external direct sum $B_1 \oplus \ldots \oplus B_r$ by the map sending $b = b_1 + \ldots + b_r$ to $(b_1, \ldots, b_r)$. We call B the *internal direct sum as algebras* of the B_i. (If $i \ne j$ and $b_i \in B_i, b_j \in B_j$, then we must have $b_i b_j \in B_i \cap B_j = 0$ since B_i and B_j are ideals; therefore, the product in B of $b_1 + \ldots + b_r$ and $b_1' + \ldots + b_r'$ is $b_1 b_1' + \ldots + b_r b_r'$.)

LEMMA 9. Let $B = B_1 \oplus \ldots \oplus B_n$ be a direct sum of algebras. Then the two-sided ideals of B are exactly the sets of the form $J_1 \oplus \ldots \oplus J_n$, where J_i is a two-sided ideal of B_i for each i.

PROOF. Let J be a two-sided ideal of B, and let $J_i = J \cap B_i$ for each i; clearly $\oplus_{i=1}^n J_i \subseteq J$. Let $b \in J$; then $b = b_1 + \ldots + b_n$, where $b_i \in B_i$ for each i. Fix some i, and let e_i be the element of B whose only non-zero entry is the identity of B_i; then $b_i = be_i \in J \cap B_i = J_i$. Therefore $b \in \oplus_{i=1}^n J_i$, which shows that J has the desired form. The converse is easy. ■

THEOREM 10. Let $r \in \mathbb{N}$. For each $1 \le i \le r$, let D_i be a division algebra over F, let $n_i \in \mathbb{N}$, and let $B_i = \mathcal{M}_{n_i}(D_i)$. Let B be the external direct sum of the B_i. Then B is a semisimple algebra having exactly r isomorphism classes of simple modules and exactly 2^r two-sided ideals, namely every sum of the form $\bigoplus_{j \in J} B_j$, where J is a subset of $\{1, \ldots, n\}$.

PROOF. For each i, we can write $B_i = C_{i1} \oplus \ldots \oplus C_{in_i}$ by Theorem 6, where the C_{ij} are mutually isomorphic simple B_i-modules. As noted above, each C_{ij} is also simple as a B-module. Therefore, we have $B \cong \oplus_{i,j} C_{ij}$ as B-modules, and hence B is a semisimple algebra by Lemma 3. It now follows from Proposition 4 that any simple B-module is isomorphic with some C_{ij}; but $C_{ij} \cong C_{kl}$ as B-modules iff $i = k$, so there are exactly r isomorphism classes of simple B-modules. The statement about two-sided ideals of B is an easy consequence of Theorem 8 and Lemma 9. ∎

We have just shown that a direct sum of matrix algebras over division algebras is semisimple. We will shortly prove an important theorem due to Wedderburn which asserts that the converse is also true: Any semisimple algebra is isomorphic with a direct sum of matrix algebras over division algebras. We develop the proof through a series of lemmas after first introducing some new concepts.

If M is an A-module, then composition of mappings gives a multiplication in $\text{End}_A(M)$, and hence $\text{End}_A(M)$ is an F-algebra. We call $\text{End}_A(M)$ the *endomorphism algebra* of M.

We define the *opposite algebra* B^{op} of an algebra B to be the set B endowed with the usual addition and scalar multiplication but the opposite multiplication. Given $a, b \in B$, we shall use ab to denote their product in B and $a \cdot b$ to denote their product in B^{op}, so that $a \cdot b = ba$ by definition. Observe that $(B^{op})^{op} \cong B$. If B is a division algebra, then so is B^{op}. The opposite of a direct sum of algebras is the direct sum of the opposite algebras, since multiplication in the direct sum is defined componentwise.

Endomorphism algebras and opposite algebras are closely related:

LEMMA 11. Let B be an algebra. Then $B^{op} \cong \text{End}_B(B)$.

PROOF. Let $\varphi \in \text{End}_B(B)$, and let $a = \varphi(1)$. Then we have $\varphi(b) = b\varphi(1) = ba$ for any $b \in B$, and hence φ is equal to the endomorphism ρ_a given by right multiplication by a. Therefore, we have $\text{End}_B(B) = \{\rho_a \mid a \in B\}$, and so $\text{End}_B(B)$ and B are in bijective correspondence. To finish the proof, it suffices to show that $\rho_a \rho_b = \rho_{a \cdot b}$ for any $a, b \in B$. Let $a, b, x \in B$; then we have $(\rho_a \rho_b)(x) = \rho_a(xb) = xba = \rho_{ba}(x) = \rho_{a \cdot b}(x)$ as required. ∎

The above result suggests that we can gain information about semisimple algebras by studying the properties of the endomorphism algebras of semisimple modules; the next result indicates that this can be accomplished by looking at the special case of modules that are direct sums of a single simple module.

LEMMA 12. Let $S_1, \ldots, S_r$ be the distinct simple A-modules; for each i, let U_i be a direct sum of copies of S_i, and let $U = U_1 \oplus \ldots \oplus U_r$. Then we have $\mathrm{End}_A(U) \cong \mathrm{End}_A(U_1) \oplus \ldots \oplus \mathrm{End}_A(U_r)$.

PROOF. Let $\varphi \in \mathrm{End}_A(U)$, and fix some i. Every composition factor of U_i is isomorphic with S_i, so by the Jordan-Hölder theorem for modules we see that the same is true of $\varphi(U_i)$, since $\varphi(U_i)$ is isomorphic with a quotient of U_i. Suppose that $\varphi(U_i)$ were not contained in U_i. Then the image of $\varphi(U_i)$ in U/U_i under the natural map would be a non-zero submodule having S_i as a composition factor. But it follows from the hypothesis that the composition factors of U/U_i are exactly those S_j for $j \neq i$, and hence a submodule of U/U_i cannot have S_i as a composition factor. Therefore for each i, we can define $\varphi_i \in \mathrm{End}_A(U_i)$ to be the restriction to U_i of φ. In this way, we define a map $\Gamma \colon \mathrm{End}_A(U) \to \mathrm{End}_A(U_1) \oplus \ldots \oplus \mathrm{End}_A(U_r)$ by setting $\Gamma(\varphi) = (\varphi_1, \ldots, \varphi_r)$. We find that Γ is an A-module monomorphism. Now let $(\varphi_1, \ldots, \varphi_r) \in \mathrm{End}_A(U_1) \oplus \ldots \oplus \mathrm{End}_A(U_r)$. We define $\hat{\varphi} \in \mathrm{End}_A(U)$ as follows: Given $x \in U$, we write $x = x_1 + \ldots + x_r$ where $x_i \in U_i$ for each i, and we define $\hat{\varphi}(x) = \varphi_1(x_1) + \ldots + \varphi_r(x_r)$. We have $(\varphi_1, \ldots, \varphi_r) = \Gamma(\hat{\varphi})$, which shows that Γ is surjective. ∎

LEMMA 13. If S is a simple A-module, then for any $n \in \mathbb{N}$ we have $\mathrm{End}_A(nS) \cong \mathcal{M}_n(\mathrm{End}_A(S))$.

PROOF. We regard the elements of nS as being column vectors of length n with entries from S. Let $\Phi = (\varphi_{ij}) \in \mathcal{M}_n(\mathrm{End}_A(S))$. We define $\Gamma(\Phi) \colon nS \to nS$ by

$$\Gamma(\Phi) \begin{pmatrix} s_1 \\ \vdots \\ s_n \end{pmatrix} = \begin{pmatrix} \varphi_{11} & \cdots & \varphi_{1n} \\ \vdots & \ddots & \vdots \\ \varphi_{n1} & \cdots & \varphi_{nn} \end{pmatrix} \begin{pmatrix} s_1 \\ \vdots \\ s_n \end{pmatrix} = \begin{pmatrix} \varphi_{11}(s_1) + \ldots + \varphi_{1n}(s_n) \\ \vdots \\ \varphi_{n1}(s_1) + \ldots + \varphi_{nn}(s_n) \end{pmatrix}.$$

We find that $\Gamma(\Phi)(a\mathbf{s} + \mathbf{t}) = a[\Gamma(\Phi)(\mathbf{s})] + \Gamma(\Phi)(\mathbf{t})$ for any $a \in A$ and $\mathbf{s}, \mathbf{t} \in nS$ since each φ_{ij} is an A-module homomorphism; hence $\Gamma(\Phi) \in \mathrm{End}_A(nS)$. We leave it to the reader to check that the map $\Gamma \colon \mathcal{M}_n(\mathrm{End}_A(S)) \to \mathrm{End}_A(nS)$ defined in this way is an algebra monomorphism. Now let ψ be an element of $\text{End}_A(nS)$. For each $1 \le i, j \le n$, we define $\psi_{ij} \colon S \to S$ implicitly by

$$
\psi \begin{pmatrix} s \\ 0 \\ \vdots \\ 0 \end{pmatrix} = \begin{pmatrix} \psi_{11}(s) \\ \psi_{21}(s) \\ \vdots \\ \psi_{n1}(s) \end{pmatrix}, \cdots, \psi \begin{pmatrix} 0 \\ \vdots \\ 0 \\ s \end{pmatrix} = \begin{pmatrix} \psi_{1n}(s) \\ \vdots \\ \psi_{(n-1)n}(s) \\ \psi_{nn}(s) \end{pmatrix}.
$$

We find that each $\psi_{ij} \in \text{End}_A(S)$. Let $\Psi = (\psi_{ij}) \in \mathcal{M}_n(\text{End}_A(S))$; then $\Gamma(\Psi) = \psi$, which shows that Γ is surjective, as required. ∎

If S is a simple A-module, then it follows immediately from Schur's lemma that $\text{End}_A(S)$ is a division algebra. If the ground field F is algebraically closed, then we can be more specific about the structure of $\text{End}_A(S)$:

LEMMA 14. Suppose that F is algebraically closed, and let S be a simple A-module. Then $\text{End}_A(S) \cong F$.

(This result is not necessary for the development of the general structure theory of semisimple algebras, but as we shall see in Chapter 6 it is critical in the application of the theory of semisimple algebras to ordinary representation theory.)

PROOF. Let $\varphi \in \text{End}_A(S)$. Viewing φ as an invertible F-linear self-map of the finite-dimensional F-vector space S, we see since F is algebraically closed that φ has a non-zero eigenvalue $\lambda_\varphi \in F$. If I is the identity element of $\text{End}_A(S)$, then $\varphi - \lambda_\varphi I \in \text{End}_A(S)$ has non-zero kernel and hence is not invertible, which forces $\varphi = \lambda_\varphi I$ since $\text{End}_A(S)$ is a division algebra. The map sending φ to λ_φ is an isomorphism from $\text{End}_A(S)$ to F. ∎

LEMMA 15. Let B be an algebra. Then $\mathcal{M}_n(B)^{op} \cong \mathcal{M}_n(B^{op})$ for any $n \in \mathbb{N}$.

PROOF. We define $\psi \colon \mathcal{M}_n(B)^{op} \to \mathcal{M}_n(B^{op})$ by letting $\psi(X)$ be the transpose X^t of the matrix X; this map ψ is clearly bijective. Let $X = (x_{ij})$ and $Y = (y_{ij})$ be elements of $(\mathcal{M}_n(B))^{op}$. Then for any i and j we have

$$
(\psi(X)\psi(Y))_{ij} = \sum_{k=1}^{n} \psi(X)_{ik} \cdot \psi(Y)_{kj} = \sum_{k=1}^{n} X_{ik}^{t} \cdot Y_{kj}^{t}
$$
$$
= \sum_{k=1}^{n} x_{ki} \cdot y_{jk} = \sum_{k=1}^{n} y_{jk} x_{ki}
$$
$$
= (YX)_{ji} = (YX)_{ij}^{t} = (X \cdot Y)_{ij}^{t}
$$
$$
= \psi(X \cdot Y)_{ij}
$$

and hence $\psi(X \cdot Y) = \psi(X)\psi(Y)$; from this, we see that ψ is an algebra homomorphism, which completes the proof. ∎

Putting the pieces together, we now have Wedderburn's main structure theorem:

THEOREM 16. The algebra A is semisimple iff it is isomorphic with a direct sum of matrix algebras over division algebras.

PROOF. Suppose that the algebra A is semisimple. Then we can write A in the form $A = U_1 \oplus \ldots \oplus U_r$, where each U_i is the direct sum of n_i copies of a simple A-module S_i, and no two of the S_i are isomorphic. We have

$$
\begin{aligned}
A^{op} &\cong \mathrm{End}_A(A) && \text{by Lemma 11} \\
&\cong \mathrm{End}_A(U_1) \oplus \ldots \oplus \mathrm{End}_A(U_r) && \text{by Lemma 12} \\
&\cong \mathrm{End}_A(n_1 S_1) \oplus \ldots \oplus \mathrm{End}_A(n_r S_r) \\
&\cong \mathcal{M}_{n_1}(\mathrm{End}_A(S_1)) \oplus \ldots \oplus \mathcal{M}_{n_r}(\mathrm{End}_A(S_r)) && \text{by Lemma 13,}
\end{aligned}
$$

and hence

$$
\begin{aligned}
A &\cong [\mathcal{M}_{n_1}(\mathrm{End}_A(S_1)) \oplus \ldots \oplus \mathcal{M}_{n_r}(\mathrm{End}_A(S_r))]^{op} \\
&\cong \mathcal{M}_{n_1}(\mathrm{End}_A(S_1))^{op} \oplus \ldots \oplus \mathcal{M}_{n_r}(\mathrm{End}_A(S_r))^{op} \\
&\cong \mathcal{M}_{n_1}(\mathrm{End}_A(S_1)^{op}) \oplus \ldots \oplus \mathcal{M}_{n_r}(\mathrm{End}_A(S_r)^{op}) \quad \text{by Lemma 15.}
\end{aligned}
$$

Since the endomorphism algebra of a simple module is a division algebra, and since the opposite algebra of a division algebra is also a division algebra, we now see that any semisimple algebra is isomorphic with a direct sum of matrix algebras over division algebras. The converse was established in Theorem 10. ∎

As an immediate corollary, we have another celebrated result of Wedderburn:

THEOREM 17. The algebra A is simple iff it is isomorphic with a matrix algebra over a division algebra.

PROOF. Suppose that A is simple. Then A is semisimple by Lemma 7, so by Theorem 16 A is isomorphic with a direct sum of r matrix algebras over division F-algebras, and hence by Theorem 10 A has exactly 2^r ideals. But A is simple and thus has exactly 2 ideals, so we must have $r = 1$; hence any simple algebra is isomorphic with a matrix algebra over a division algebra. The converse was established in Theorem 8. ∎

We now see, albeit indirectly, that an algebra is semisimple iff it is a direct sum of simple algebras, which affirms the consistency of our choices of terminology.

Semisimple algebras over algebraically closed fields have a more specific classification than those over arbitrary fields:

THEOREM 18. Suppose that the field F is algebraically closed. Then any semisimple algebra is isomorphic with a direct sum of matrix algebras over F.

PROOF. This follows from Lemma 14 and the proof of Theorem 16. ∎

In the remainder of this section, A is an algebra as before, but now possibly without unit. An element $x \in A$ is called *nilpotent* if $x^n = 0$ for some $n \in \mathbb{N}$. For example, the matrix $\left(\begin{smallmatrix} 0 & 1 \\ 0 & 0 \end{smallmatrix} \right)$ is a nilpotent element of $\mathcal{M}_2(F)$ since its square is the zero matrix. More generally, any upper triangular element of $\mathcal{M}_n(F)$ whose main diagonal consists solely of zeroes is nilpotent. An ideal I of A is said to be *nilpotent* if $I^n = 0$ for some $n \in \mathbb{N}$, where I^n is the ideal spanned by all products of n elements of I. (By ideal, we shall always in this section mean two-sided ideal.) Taking $I = A$, we say that the algebra A is nilpotent if there is some $n \in \mathbb{N}$ such that any product of n elements of A equals zero; in particular, every element of a nilpotent algebra is nilpotent, and consequently a nilpotent algebra cannot be an algebra with unit.

LEMMA 19. Let I and J be nilpotent ideals of A. Then $I + J$ is also a nilpotent ideal of A.

PROOF. Let $m, n \in \mathbb{N}$ be such that $I^m = J^n = 0$. The product of any $m + n$ elements of $I + J$ can be written as a sum of terms of the form $z_1 \cdots z_{m+n}$, where each z_i lies in either I or J. In any such term, by the pigeonhole principle either at least m of the z_i lie in I or at least n lie in J. In the former case, since I is an ideal we can rewrite $z_1 \cdots z_{m+n}$ as a product of m elements of I, and hence we have $z_1 \cdots z_{m+n} = 0$ by the nilpotence of I; the other case is similar. Therefore, the product of any $m + n$ elements of $I + J$ is zero, and hence $(I + J)^{m+n} = 0$. ∎

COROLLARY 20. A has a largest nilpotent ideal.

(What we mean by this is that there is a nilpotent ideal of A that contains every nilpotent ideal of A.)

PROOF. The sum of all nilpotent ideals of A can be realized as a finite sum since A is finite-dimensional; this finite sum contains all nilpotent ideals of A, and we see that it is nilpotent by applying Lemma 19 and an induction argument. ∎

LEMMA 21. Let M and N be submodules of A such that A/M and A/N are semisimple A-modules. Then $A/M \cap N$ is a semisimple A-module.

PROOF. The map from $A/M \cap N$ to $A/M \oplus A/N$ that sends $a + (M \cap N)$ to $(a + M, a + N)$ is easily seen to be an A-module monomorphism; the result now follows from Lemma 2. ∎

COROLLARY 22. A has a smallest submodule having semisimple quotient.

(Similarly, what we mean by this is that there is a submodule of A having semisimple quotient that is contained in every submodule of A that has semisimple quotient.)

PROOF. As in Corollary 20, the intersection of all submodules of A having semisimple quotient can be realized as a finite intersection of such submodules, and we see via Lemma 21 and an induction argument that this finite intersection has semisimple quotient. ∎

Let M be an A-module. We define $\mathrm{Ann}(M) = \{a \in A \mid aM = 0\}$; this is called the *annihilator* of M, and we see easily that it is an ideal of A. Given a family $\{M_t\}_{t \in T}$ of A-modules, we define their *common annihilator* to be $\bigcap_{t \in T} \mathrm{Ann}(M_t)$.

The following result relates the concepts introduced in the previous paragraphs. The key parts of this theorem are due to Wedderburn, although not in this formulation.

THEOREM 23. Let A be an algebra with unit. Then the following are equal:

- The largest nilpotent ideal of A.
- The common annihilator of all simple A-modules.
- The smallest submodule of A having semisimple quotient.

This object is called the *radical* of A and will be denoted $\mathrm{rad}(A)$. It is often called the Jacobson radical (and then denoted $J(A)$) after Nathan Jacobson, a student of Wedderburn.

PROOF. Let I be the largest nilpotent ideal of A, which exists by Corollary 20; let $\mathcal{A}$ be the common annihilator of all simple A-modules; and let M be the smallest submodule of A having semisimple quotient, which exists by Corollary 22.

Let J be a nilpotent ideal of A, and let S be a simple A-module. As JS is a submodule of S, either $JS = 0$ or $JS = S$. If $JS = S$, then we have $J^k S = S$ for all $k \in \mathbb{N}$; but $J^n = 0$ for some $n \in \mathbb{N}$ since J is nilpotent, so this is a contradiction. Therefore $JS = 0$; thus every nilpotent ideal annihilates every simple A-module, and so it follows that $I \subseteq \mathcal{A}$.

Since A is finite-dimensional, it has as an A-module a composition series $A = A_0 \supset A_1 \supset \ldots \supset A_r = 0$. Each successive quotient A_i/A_{i+1} is a simple A-module, and hence $\mathcal{A}(A_i/A_{i+1}) = 0$, or equivalently $\mathcal{A}A_i \subseteq A_{i+1}$. Therefore, $\mathcal{A}^k \subseteq A_k$ for every $1 \leq k \leq r$, and in particular $\mathcal{A}^r \subseteq A_r = 0$, which gives $\mathcal{A} \subseteq I$. Therefore $I = \mathcal{A}$.

Now $\mathcal{A}$ annihilates all simple A-modules, so in addition it annihilates all semisimple A-modules. Since A/M is semisimple, we have $\mathcal{A}(A/M) = 0$, and hence $\mathcal{A}A \subseteq M$; as $\mathcal{A} \subseteq \mathcal{A}A$ since A has unit, this gives $\mathcal{A} \subseteq M$.

Suppose that $\mathcal{A} \subset M$. Then there is a simple A-module S such that $MS \neq 0$, which forces $MS = S$ since MS is a submodule of S. Moreover, Ms is a submodule of S for every $s \in S$, so there must be some $s \in S$ such that $Ms = S$. Let $m \in M$ be such that $ms = -s$, in which case $m + 1$ lies in the annihilator of s. The annihilator of s is a proper submodule of A, and since A is finite-dimensional it follows that the annihilator of s, and in particular $m + 1$, lies in some maximal submodule N of A. Since N is maximal, A/N is simple and hence semisimple, and so $M \subseteq N$. But now we have $m \in N$, which gives $1 = (m + 1) - m \in N$ and hence $A = N$; this is a contradiction since $N \subset A$. Therefore, we must have $A = M$. ∎

COROLLARY 24. Let A be an algebra with unit. Then A is semisimple iff A has no non-zero nilpotent ideals.

PROOF. By Lemma 3, A is semisimple iff A is semisimple as an A-module, and by Theorem 23 this is true iff $\mathrm{rad}(A) = 0$; but we also see from Theorem 23 that $\mathrm{rad}(A) = 0$ iff A has no non-zero nilpotent ideals. ∎

We have just seen that an algebra with unit whose radical is zero is semisimple. This suggests the following question: What can be said about an algebra without unit whose radical is zero? This question has an answer that is perhaps surprising: There are no such algebras. Before establishing this, we need a lemma that enables us to imbed an algebra without unit inside an algebra with unit.

LEMMA 25. Let A be an algebra, possibly without unit. Then there exists an algebra B with unit and an ideal I of B such that, as algebras, we have $I \cong A$ and $B/I \cong F$.

PROOF. Let B be the set $F \times A$, endowed with componentwise addition and scalar multiplication and with the following multiplication: We define $(\lambda, a)(\mu, b) = (\lambda\mu, \lambda b + \mu a + ab)$ for $\lambda, \mu \in F$ and $a, b \in A$. Let $I = \{(0, a) \mid a \in A\} \subseteq B$. We leave it to the reader to verify that B is an algebra with unit $(1, 0)$ satisfying the stated conditions. ∎

THEOREM 26. Let A be a non-zero algebra, possibly without unit. If A has no non-zero nilpotent ideals, then A is an algebra with unit.

PROOF. Let B and I be as in Lemma 25, and let J be a nilpotent ideal of B. Then $J \cap I$ is a nilpotent ideal that is contained in I; but $I \cong A$ by Lemma 25, so from the hypothesis we deduce that $J \cap I = 0$. The first isomorphism theorem for modules now gives $J = J/J \cap I \cong I + J/I$; thus, $I + J/I$ is a nilpotent ideal in B/I.

But $B/I \cong F$ by Lemma 25, and F has no non-zero proper ideals, so we must have $J = 0$. Therefore, B has no nilpotent ideals and hence is semisimple by Corollary 24; it now follows from Theorem 16 that B is isomorphic with a direct sum of matrix algebras over division algebras. By Theorem 10, every ideal of B is also a direct sum of matrix algebras over division algebras, and thus every non-zero ideal of B is an algebra with unit. But $A \cong I$ is a non-zero ideal of B by Lemma 25, and therefore A is an algebra with unit. ∎

We can now prove a final theorem of Wedderburn; this result will allow us to complete the proof of Kolchin's theorem, which was left unfinished in Section 5.

THEOREM 27. Suppose that the algebra A can be generated as an F-vector space by a set consisting of nilpotent elements. Then A is nilpotent.

PROOF. Let $\bar{F}$ be an algebraically closed field that contains F. (Such fields certainly exist; see [18, Section 8.1].) Let $\bar{A} = \bar{F} \otimes_F A$. Then $\bar{A}$ is an $\bar{F}$-vector space with $\dim_{\bar{F}} \bar{A} = \dim_F A$, and any F-basis for A induces an $\bar{F}$-basis for $\bar{A}$. Moreover, $\bar{A}$ has an algebra structure that extends that of A. We can view A as a subalgebra of $\bar{A}$ via the inclusion of F in $\bar{F}$ and the isomorphism (as in Proposition 12.2) of A with $F \otimes_F A$. (The algebra $\bar{A}$ is called the *extension of scalars* from F to $\bar{F}$ of A.) By hypothesis, A can be generated as a vector space by a set consisting of nilpotent elements; hence $\bar{A}$ also has this property. Since $A \subseteq \bar{A}$, it suffices to prove that $\bar{A}$ is nilpotent. Therefore, without loss of generality, we can assume that F is algebraically closed.

We shall use induction on $\dim_F A$. If $\dim_F A = 1$, then every element of A is a scalar multiple of a nilpotent element, and hence A is nilpotent. Thus we assume that $\dim_F A > 1$. Suppose that A has a non-zero nilpotent ideal I. Then $\dim_F A/I < \dim_F A$, and so by induction A/I, and consequently A, is nilpotent. It remains only to consider the case where A has no non-zero nilpotent ideals. In this case, A is an algebra with unit by Theorem 26 and hence is semisimple by Corollary 24. We now see from Theorem 18 that A is isomorphic with a direct sum of matrix algebras over F. Let $\mathcal{M}_n(F)$ be one of these summands. Since A can be generated as a vector space by nilpotent elements, we see that the same is true of $\mathcal{M}_n(F)$. If $M \in \mathcal{M}_n(F)$, then since F is algebraically closed it follows via

Jordan form that M is similar to an upper triangular matrix; if M is nilpotent, then this observation implies that the trace of M must be zero. Therefore, the nilpotent elements that generate $\mathcal{M}_n(F)$ are contained in a proper subspace of $\mathcal{M}_n(F)$, namely the kernel of the trace map; this is a contradiction, so this case cannot arise. ∎

We now give the application of this theorem used in Section 5.

PROPOSITION 28. Let $V_n(F)$ be the space of column vectors of length $n \in \mathbb{N}$ with entries from a field F, and let H be a unipotent subgroup of $\mathrm{GL}(n, F)$. Then there exists some $0 \neq \mathbf{v} \in V_n(F)$ such that $x\mathbf{v} = \mathbf{v}$ for all $x \in H$.

PROOF. Let V be the subspace of $\mathrm{GL}(n, F)$ spanned by the set $S = \{x - I \mid x \in H\}$, where I denotes the identity matrix. If $x \in H$, then we have $(x - I)^n = 0$ since x is unipotent, and consequently S consists of nilpotent elements. Now for $x, y \in H$, we observe that $(x - I)(y - I) = (xy - I) - (x - I) - (y - I) \in V$. Thus, V is closed under matrix multiplication, and we conclude that V is an algebra. Since V is generated as a vector space by the set S of nilpotent elements, V is nilpotent by Theorem 27. In particular, there is some minimal $t \in \mathbb{N}$ such that $(x_1 - I) \cdots (x_t - I) = 0$ for any $x_1, \ldots, x_t \in H$. By the minimality of t, there must exist some $y_1, \ldots, y_{t-1} \in H$ such that $(y_1 - I) \cdots (y_{t-1} - I)$ is not the zero matrix, and thus there must exist non-zero vectors $\mathbf{v}$ and $\mathbf{v}'$ in $V_n(F)$ such that $(y_1 - I) \cdots (y_{t-1} - I)\mathbf{v}' = \mathbf{v}$. Now for any $x \in H$ we have $(x - I)\mathbf{v} = (x - I)(y_1 - I) \cdots (y_{t-1} - I)\mathbf{v}' = 0\mathbf{v}' = 0$ and hence $x\mathbf{v} = \mathbf{v}$. ∎

EXERCISES

Throughout these exercises, A denotes a finite-dimensional algebra with unit over a field F, and all A-modules are finitely generated.

1. Let $n \in \mathbb{N}$. Let V be the n-dimensional subspace of A^n spanned by the identity elements of the summands. Show that if M is an A-module and $T: V \to M$ is a linear transformation, then there is a unique A-module homomorphism from A^n to M which extends T.

2. (cont.) Suppose that B is an A-module having an n-dimensional subspace U such that whenever M is an A-module and $T: U \to M$ is a linear transformation, there is a unique extension of T to an A-module homomorphism from B to M. Show that B and A^n

are isomorphic A-modules. (Modules of the form A^n for some $n \in \mathbb{N}$ are called *free* A-modules; hence this exercise provides an alternate characterization of freeness, which we shall generalize for group algebras in the exercises to Section 16.)

3. Let $n \in \mathbb{N}$, and let $T_n(F)$ be the algebra of upper triangular $n \times n$ matrices. Show that the set $V_n(F)$ of column vectors over F of length n is a $T_n(F)$-module that has a unique composition series in which every simple $T_n(F)$-module appears exactly once as a composition factor.

4. (cont.) Show that the $T_n(F)$-module $T_n(F)$ is isomorphic with the direct sum of all non-zero submodules of $V_n(F)$.

5. (cont.) Find the largest nilpotent ideal of $T_n(F)$, and show without reference to Theorem 23 that it is also the common annihilator of all simple $T_n(F)$-modules and the smallest submodule of $T_n(F)$ having semisimple quotient.

6. Let U be an A-module, let $n \in \mathbb{N}$, and let U^n be the set of column vectors of length n with entries from U, considered in the obvious way as an $\mathcal{M}_n(A)$-module. Show that U is a simple A-module iff U^n is a simple $\mathcal{M}_n(A)$-module.

7. (cont.) Show that $\operatorname{Hom}_A(U, V) \cong \operatorname{Hom}_{\mathcal{M}_n(A)}(U^n, V^n)$ for any A-modules U and V.

8. (cont.) Show that every $\mathcal{M}_n(A)$-module is isomorphic with a module of the form U^n for some A-module U. (What Exercises 6–8 show is that the module theory of the algebras A and $\mathcal{M}_n(A)$ is, in a sense that can be made precise, the same. In the language of category theory, we say that the categories of A-modules and $\mathcal{M}_n(A)$-modules are Morita equivalent.)

9. Show that A is a simple A-module iff A is a division algebra.

10. We can clearly regard an $A/\operatorname{rad} A$-module as being an A-module; on the other hand, a simple A-module is annihilated by $\operatorname{rad}(A)$ and hence can be considered as an $A/\operatorname{rad} A$-module. Show that an A-module is simple iff it is a simple $A/\operatorname{rad}(A)$-module. (Since $A/\operatorname{rad}(A)$ is a semisimple algebra, this implies that the determination of simple modules over arbitrary algebras reduces to the case of semisimple algebras.)

11. Let B be a finite-dimensional F-algebra, possibly without unit. Show that if B is a simple algebra whose multiplication is not identically zero, then B is isomorphic with a matrix algebra over a division algebra.

FURTHER EXERCISES

Let A be as above, and let M be a finitely generated A-module.

12. Show that the following objects (exist and) are equal: the intersection of all maximal submodules of M; the smallest submodule of M having semisimple quotient; and $\mathrm{rad}(A)M$. This submodule of M is called the *radical* of M and is denoted $\mathrm{rad}(M)$. (Observe that our two definitions of the symbol $\mathrm{rad}(A)$ agree.)

We define a descending series $\mathrm{rad}^n(M)$ of submodules of M by setting $\mathrm{rad}^0(M) = M$ and $\mathrm{rad}^n(M) = \mathrm{rad}(\mathrm{rad}^{n-1}(M))$ for $n \in \mathbb{N}$. It follows from Exercise 12 that we have $\mathrm{rad}^n(M) = (\mathrm{rad}(A))^n M$ for any $n \in \mathbb{N}$. This series is called the *radical series* of M. Observe that $\mathrm{rad}^n(M) = 0$ for some n, since $\mathrm{rad}(A)$ is a nilpotent ideal, and that each successive quotient of the radical series is a semisimple module, with M being semisimple iff $\mathrm{rad}(M) = 0$.

13. (cont.) Show that M has a unique largest semisimple submodule and that this submodule is equal to $\{m \in M \mid \mathrm{rad}(A)m = 0\}$. We call this submodule the *socle* of M, and we denote it by $\mathrm{soc}(M)$.

14. (cont.) We define an ascending series $\mathrm{soc}^n(M)$ of submodules of M as follows: We let $\mathrm{soc}^0(M) = 0$ and $\mathrm{soc}^1(M) = \mathrm{soc}(M)$, and for $n > 1$ we let $\mathrm{soc}^n(M)$ be the submodule of M such that $\mathrm{soc}^n(M)/\mathrm{soc}^{n-1}(M) = \mathrm{soc}(M/\mathrm{soc}^{n-1}(M))$. Show that we have $\mathrm{soc}^n(M) = \{m \in M \mid \mathrm{rad}(A)^n m = 0\}$ for any $n \in \mathbb{N}$. (This series is called the *socle series* of M. As with the radical series, we observe that $\mathrm{soc}^n(M) = M$ for some n, since $\mathrm{rad}(A)$ is nilpotent, and that each successive quotient of the socle series is semisimple, with M being semisimple iff $\mathrm{soc}(M) = M$.)

15. (cont.) Show that $\mathrm{rad}^n(M) = 0$ iff $\mathrm{soc}^n(M) = M$; conclude that the radical and socle series of M have the same finite length. If this common length is r, then show that $\mathrm{rad}^n(M) \subseteq \mathrm{soc}^{r-n}(M)$ for every $0 \le n \le r$.

6
Group Representations

In this final chapter, we explore the character theory of finite groups. Section 14 introduces characters and develops some of their elementary properties. This development is continued in Section 15 with the introduction of the character table and the verification of the orthogonality relations. The bulk of this section is devoted to a series of instructive examples involving the computation of character tables. In Section 16, we momentarily return to a more general setting with the definition of induced modules; we then study induced characters, finishing with a well-known group-theoretic theorem of Frobenius that (at present) cannot be established without invoking character theory.

14. Characters

In this chapter G denotes a finite group, and all $\mathbb{C}G$-modules are finitely generated (or equivalently have finite dimension as $\mathbb{C}$-vector spaces), where $\mathbb{C}$ is the field of complex numbers.

Since $\mathbb{C}$ is an algebraically closed field of characteristic zero, we obtain from Wedderburn theory specific information about the nature of the algebra $\mathbb{C}G$:

THEOREM 1. There is some $r \in \mathbb{N}$ and some $f_1, \ldots, f_r \in \mathbb{N}$ such that $\mathbb{C}G \cong \mathcal{M}_{f_1}(\mathbb{C}) \oplus \ldots \oplus \mathcal{M}_{f_r}(\mathbb{C})$ as $\mathbb{C}$-algebras. Furthermore, there are exactly r isomorphism classes of simple $\mathbb{C}G$-modules, and if we let $S_1, \ldots, S_r$ be representatives of these r classes, then we can order the S_i so that $\mathbb{C}G \cong f_1 S_1 \oplus \ldots \oplus f_r S_r$ as $\mathbb{C}G$-modules, where $\dim_{\mathbb{C}} S_i = f_i$ for each i. Any $\mathbb{C}G$-module can be written uniquely in the form $a_1 S_1 \oplus \ldots \oplus a_r S_r$, where the a_i are non-negative integers.

PROOF. The first statement follows from Corollary 12.8 and Theorem 13.18; the second, from Theorems 13.6 and 13.10, where we take S_i to be the space of column vectors of length f_i with the canonical module structure over the ith summand $\mathcal{M}_{f_i}(\mathbb{C})$; and the third, from Proposition 13.5. ∎

The $\mathbb{C}$-dimensions $f_1, \ldots, f_r$ of the r simple $\mathbb{C}G$-modules are called the *degrees* of G. The trivial $\mathbb{C}G$-module $\mathbb{C}$ is one-dimensional and hence simple, so G will always have at least one of its degrees equal to 1, and by convention we will set $f_1 = 1$.

COROLLARY 2. We have $\sum_{i=1}^{r} f_i^2 = |G|$.

PROOF. Theorem 1 gives

$$|G| = \dim_{\mathbb{C}} \mathbb{C}G = \dim_{\mathbb{C}} \left(\bigoplus_{i=1}^{r} \mathcal{M}_{f_i}(\mathbb{C}) \right) = \sum_{i=1}^{r} \dim_{\mathbb{C}} \mathcal{M}_{f_1}(\mathbb{C})$$

$$= \sum_{i=1}^{r} f_i^2. \quad \blacksquare$$

In fact, the degrees of G must divide $|G|$; we shall not use this fact, but we provide a proof of it in the Appendix.

We now establish an important connection between the number of simple $\mathbb{C}G$-modules and the structure of G.

THEOREM 3. The number r of simple $\mathbb{C}G$-modules is equal to the number of conjugacy classes of G.

PROOF. Let Z be the center of $\mathbb{C}G$, meaning that Z is the subalgebra of $\mathbb{C}G$ consisting of all elements that commute with every element of $\mathbb{C}G$. Using Theorem 1, we observe that Z is isomorphic with the center of $\mathcal{M}_{f_1}(\mathbb{C}) \oplus \ldots \oplus \mathcal{M}_{f_r}(\mathbb{C})$ and hence is isomorphic with the direct sum of the centers of the $\mathcal{M}_{f_i}(\mathbb{C})$. As noted in the proof of Proposition 6.4, the center of any $\mathcal{M}_{f_i}(\mathbb{C})$ consists only of the scalar matrices and hence is isomorphic with $\mathbb{C}$; thus $Z \cong \mathbb{C}^r$, and in particular $\dim_{\mathbb{C}} Z = r$.

Consider an element $\sum_{g \in G} \lambda_g g$ of Z. For any $h \in G$, we have $(\sum_g \lambda_g g)h = h(\sum_g \lambda_g g)$, giving $\sum_g \lambda_g g = \sum_g \lambda_g h^{-1} g h = \sum_g \lambda_{hgh^{-1}} g$. Therefore, we have $\lambda_g = \lambda_{hgh^{-1}}$ for every $g, h \in G$, and so we conclude that the coefficients of elements of Z are constant on conjugacy classes. It follows that a basis for Z is the set of *class sums*, which are the sums of the form $\sum_{g \in K} g$ where K is a conjugacy class of G. Thus, $\dim_{\mathbb{C}} Z$ is equal to the number of conjugacy classes of G, which completes the proof. ∎

We should mention that, in general, there is no natural bijective correspondence between the conjugacy classes of G and the simple $\mathbb{C}G$-modules. However, if G is a symmetric group, then there is such a correspondence, although we shall not develop it here; for an overview, see [13, Section 4.1].

If U is a $\mathbb{C}G$-module, then each $g \in G$ defines an invertible linear transformation of U that sends $u \in U$ to gu. We define the *character* of U to be the function $\chi_U \colon G \to \mathbb{C}$, where $\chi_U(g)$ is the trace of this linear transformation of U defined by g. For example, $\chi_U(1) = \dim_{\mathbb{C}} U$, since the identity element of G induces the identity transformation on U. If $\rho \colon G \to \mathrm{GL}(U)$ is the representation corresponding to U, then $\chi_U(g)$ is just the trace of the map $\rho(g)$. Isomorphic $\mathbb{C}G$-modules have equal characters. We observe that for any $g, h \in G$, the linear transformations of U defined by g and hgh^{-1} are similar and hence have the same trace. Therefore, any character of G is constant on each conjugacy class of G, meaning that the value of the character on any two conjugate elements is the same.

For example, let $U = \mathbb{C}G$, and let $g \in G$. By considering the matrix of the linear transformation defined by g with respect to the basis G of $\mathbb{C}G$, we see that $\chi_U(g)$ is equal to the number of elements $x \in G$ for which $gx = x$. Therefore, we have $\chi_U(1) = |G|$ and $\chi_U(g) = 0$ for every $1 \ne g \in G$. This character is called the *regular character* of G.

The theory of characters was developed by Frobenius and others, starting in 1896. However, at first nothing was known about linear representations, let alone modules over the group algebra; Frobenius defined characters as being functions from G to $\mathbb{C}$ satisfying certain properties, but it turned out that his characters were exactly the trace functions of finitely generated $\mathbb{C}G$-modules.

We will denote the characters of the r simple $\mathbb{C}G$-modules by

$\chi_1, \ldots, \chi_r$; these characters will be referred to as the *irreducible characters* of G. Whenever we say that $S_1, \ldots, S_r$ are the distinct simple $\mathbb{C}G$-modules, we implicitly order them so that $\chi_{S_i} = \chi_i$ for each i. In keeping with the convention that f_1 denotes the degree of the trivial representation, we let χ_1 be the character of the trivial representation; we call χ_1 the *principal character* of G, and we have $\chi_1(g) = 1$ for all $g \in G$.

A character of a one-dimensional $\mathbb{C}G$-module is called a *linear character*. Since one-dimensional modules are simple, we see that all linear characters are irreducible. Let χ be the linear character arising from the $\mathbb{C}G$-module U, and let $g, h \in G$. Since U is one-dimensional, for any $u \in U$ we have $gu = \chi(g)u$ and $hu = \chi(h)u$, and thus $\chi(gh)u = (gh)u = \chi(g)\chi(h)u$; therefore, χ is a homomorphism from G to the multiplicative group $\mathbb{C}^\times$ of non-zero complex numbers. On the other hand, given a homomorphism $\varphi \colon G \to \mathbb{C}^\times$, we can define a one-dimensional $\mathbb{C}G$-module U by $gu = \varphi(g)u$ for $g \in G$ and $u \in U$, and we then have $\chi_U = \varphi$. Therefore, linear characters of G are exactly the same as group homomorphisms from G to $\mathbb{C}^\times$.

Our next result compiles some basic information about characters.

PROPOSITION 4. Let U be a $\mathbb{C}G$-module, let $\rho \colon G \to \mathrm{GL}(U)$ be the representation corresponding to U, and let $g \in G$ be of order n. Then:

(i) $\rho(g)$ is diagonalizable.

(ii) $\chi_U(g)$ equals the sum (with multiplicities) of the eigenvalues of $\rho(g)$.

(iii) $\chi_U(g)$ is a sum of $\chi_U(1)$ nth roots of unity.

(iv) $\chi_U(g^{-1}) = \overline{\chi_U(g)}$. (Here $\overline{z}$ denotes the complex conjugate of $z \in \mathbb{C}$.)

(v) $|\chi_U(g)| \le \chi_U(1)$.

(vi) $\{x \in G \mid \chi_U(x) = \chi_U(1)\}$ is a normal subgroup of G.

PROOF. Since $g^n = 1$, $\rho(g)$ satisfies the polynomial $X^n - 1$. But $X^n - 1$ splits into distinct linear factors in $\mathbb{C}[X]$, and so it follows that the minimal polynomial of $\rho(g)$ does also, and hence that $\rho(g)$ is diagonalizable, proving (i). It now follows that the trace of $\rho(g)$ is the sum of its eigenvalues, proving (ii). These eigenvalues are precisely the roots of the minimal polynomial of $\rho(g)$, which divides $X^n - 1$; consequently those roots are nth roots of unity, which (since $\chi_U(1) = \dim_\mathbb{C} U$) proves (iii). We see easily that any eigenvector for

$\rho(g)$ is also an eigenvector for $\rho(g^{-1})$, with the eigenvalue for $\rho(g^{-1})$ being the inverse of the eigenvalue for $\rho(g)$. Since the eigenvalues of $\rho(g)$ are roots of unity, it follows that the eigenvalues of $\rho(g^{-1})$ are the conjugates of the eigenvalues of $\rho(g)$, and from this (iv) follows easily. (v) follows immediately from (iii). We have already seen that $\chi_U(g)$ is the sum of its $\chi_U(1)$ eigenvalues, each of which is a root of unity. If this sum equals $\chi_U(1)$, then it follows that each of those eigenvalues must be 1, in which case $\rho(g)$ must be the identity map. Conversely, if $\rho(g)$ is the identity map, then we have $\chi_U(g) = \dim_{\mathbb{C}}(U) = \chi_U(1)$; therefore $\{x \in G \mid \chi_U(x) = \chi_U(1)\} = \ker \rho \trianglelefteq G$, proving (vi). ∎

Suppose that χ and ψ are characters of G. We can define new functions $\chi + \psi$ and $\chi\psi$ from G to $\mathbb{C}$ by $(\chi + \psi)(g) = \chi(g) + \psi(g)$ and $(\chi\psi)(g) = \chi(g)\psi(g)$ for $g \in G$. These new functions obtained from characters are not, a priori, characters themselves. We can also, given a scalar $\lambda \in \mathbb{C}$, define a new function $\lambda\chi \colon G \to \mathbb{C}$ by $(\lambda\chi)(g) = \lambda\chi(g)$, and consequently we can view the characters of G as elements of a $\mathbb{C}$-vector space of functions from G to $\mathbb{C}$.

PROPOSITION 5. The irreducible characters of G are, as functions from G to $\mathbb{C}$, linearly independent over $\mathbb{C}$.

PROOF. We have $\mathbb{C}G \cong M_{f_1}(\mathbb{C}) \oplus \ldots \oplus M_{f_r}(\mathbb{C})$ by Theorem 1. Let $S_1, \ldots, S_r$ be the distinct simple $\mathbb{C}G$-modules, and for each i let e_i be the identity element of $M_{f_i}(\mathbb{C})$. Fix some i. Recall that for any $g \in G$, $\chi_i(g)$ is the trace of the linear transformation on S_i defined by $g \in G$. We linearly extend χ_i to a linear map from $\mathbb{C}G$ to $\mathbb{C}$, so that $\chi_i(a)$ for $a \in \mathbb{C}G$ is the trace of the linear transformation on S_i defined by a. We observe that the linear transformation on S_i given by e_i is the identity map, and hence that $\chi_i(e_i) = \dim_{\mathbb{C}} S_i = f_i$. Moreover, if $j \neq i$, then the linear transformation on S_j given by e_i is the zero map, and hence $\chi_j(e_i) = 0$.

Now let $\lambda_1, \ldots, \lambda_r \in \mathbb{C}$ be such that $\sum_{j=1}^r \lambda_j \chi_j = 0$. From the above we see that $0 = \sum_{j=1}^r \lambda_j \chi_j(e_i) = \lambda_i f_i$ for each i; thus $\lambda_j = 0$ for all j, proving the result. ∎

LEMMA 6. $\chi_{U \oplus V} = \chi_U + \chi_V$ for any $\mathbb{C}G$-modules U and V.

PROOF. By considering a $\mathbb{C}$-basis for $U \oplus V$ whose first $\dim_{\mathbb{C}} U$ elements form a $\mathbb{C}$-basis for $U \oplus 0$ and whose remaining elements form a $\mathbb{C}$-basis for $0 \oplus V$, we see easily that $\chi_{U \oplus V}(g) = \chi_U(g) + \chi_V(g)$ for any $g \in G$. ∎

We can now show that the characters of $\mathbb{C}G$-modules suffice to distinguish between $\mathbb{C}G$-modules:

THEOREM 7. If $S_1, \ldots, S_r$ are the distinct simple $\mathbb{C}G$-modules, then the character of the $\mathbb{C}G$-module $a_1 S_1 \oplus \ldots \oplus a_r S_r$ (where the a_i are non-negative integers) is $a_1 \chi_1 + \ldots + a_r \chi_r$. Consequently, two $\mathbb{C}G$-modules are isomorphic iff their characters are equal.

PROOF. The first assertion follows directly from Lemma 6. Now suppose that $\chi_U = \chi_V$ for some $\mathbb{C}G$-modules U and V. Since $\mathbb{C}G$ is semisimple, we can write $U \cong \oplus_i a_i S_i$ and $V \cong \oplus_i b_i S_i$, where the a_i and b_i are non-negative integers. By taking characters, we have $0 = \chi_U - \chi_V = \sum_i (a_i - b_i) \chi_i$, which by Proposition 5 forces $a_i = b_i$ for all i; hence $U \cong V$. ∎

Although each character determines a $\mathbb{C}G$-module up to isomorphism by Theorem 7, there is no generic way to construct a module from its corresponding character, and so in some sense information is lost by studying characters in lieu of modules. However, characters turn out to be an efficient means of translating information about the ordinary representation theory of G to information about G itself, and for this reason we will soon turn our attention away from $\mathbb{C}G$-modules and toward their characters.

We have seen in Section 12 that if U and V are $\mathbb{C}G$-modules, then $U \otimes_{\mathbb{C}} V$ and $\mathrm{Hom}_{\mathbb{C}}(U, V)$, which we shall write here simply as $U \otimes V$ and $\mathrm{Hom}(U, V)$, admit natural $\mathbb{C}G$-module structures. We now consider the relationship between the characters of these modules and those of U and V.

PROPOSITION 8. Let U and V be $\mathbb{C}G$-modules. Then:

  (i) $\chi_{U \otimes V} = \chi_U \chi_V$.
  (ii) $\chi_{U^*} = \overline{\chi_U}$. (Recall that $U^* = \mathrm{Hom}(U, F)$.)
  (iii) $\chi_{\mathrm{Hom}(U,V)} = \overline{\chi_U} \chi_V$.

PROOF.   (i) Let $g \in G$. By part (i) of Proposition 4, the transformation defined by the action of g on U is diagonalizable; let $\{u_1, \ldots, u_m\}$ be a basis of U consisting of eigenvectors of this transformation, with respective eigenvalues $\lambda_1, \ldots, \lambda_m$. Similarly, let $\{v_1, \ldots, v_n\}$ be a basis of V consisting of eigenvectors of the transformation defined by the action of g on V, with respective eigenvalues $\mu_1, \ldots, \mu_n$. We have $\chi_U(g) = \lambda_1 + \ldots + \lambda_m$ and $\chi_V(g) = \mu_1 + \ldots + \mu_n$ by part (ii) of Proposition 4. Now $\{u_i \otimes v_j\}_{i,j}$ is a basis for $U \otimes V$, and $g(u_i \otimes v_j) = gu_i \otimes gv_j = \lambda_i u_i \otimes \mu_j v_j = \lambda_i \mu_j (u_i \otimes v_j)$ for any i, j. Hence the basis $\{u_i \otimes v_j\}_{i,j}$ consists of eigenvectors for the transformation defined by the action of g on $U \otimes V$, and thus $\chi_{U \otimes V}(g) = \sum_{i,j} \lambda_i \mu_j = (\sum_i \lambda_i)(\sum_j \mu_j) = \chi_U(g)\chi_V(g)$.

(ii) Let $g \in G$, and as in the proof of (i) let $\{u_1, \ldots, u_m\}$ be a basis for U of eigenvectors of the transformation of U defined by g, with respective eigenvalues $\lambda_1, \ldots, \lambda_m$. Let $\{\varphi_1, \ldots, \varphi_m\}$ be the dual basis of U^*, so that $\varphi_i \colon U \to \mathbb{C}$ is defined for each i by $\varphi_i(u_j) = \delta_{ij}$ for each j. Fix some j. Now $gu_j = \lambda_j u_j$, so $g^{-1}u_j = \lambda_j^{-1} u_j$. However, we observed in the proof of Proposition 4 that λ_j is a root of unity, and consequently we have $\lambda_j^{-1} = \overline{\lambda_j}$. For any i, j we now have $(g\varphi_i)(u_j) = \varphi_i(g^{-1}u_j) = \varphi_i(\overline{\lambda_j}u_j) = \overline{\lambda_j}\delta_{ij}$, and this gives $g\varphi_i = \overline{\lambda_i}\varphi_i$ for each i; therefore, the basis $\{\varphi_1, \ldots, \varphi_m\}$ of U^* consists of eigenvectors for the transformation of U^* defined by g, with respective eigenvalues $\overline{\lambda_1}, \ldots, \overline{\lambda_m}$. Therefore $\chi_{U^*}(g) = \overline{\lambda_1} + \ldots + \overline{\lambda_m} = \overline{\lambda_1 + \ldots + \lambda_m} = \overline{\chi_U(g)}$.

(iii) By Proposition 12.7, we have $\operatorname{Hom}(U, V) \cong U^* \otimes V$, and so the result follows from (i) and (ii) above. ∎

A *virtual character* of G is a $\mathbb{Z}$-linear combination of the irreducible characters of G. (Some authors prefer the term "generalized character.") Characters are virtual characters by Theorem 7.

COROLLARY 9. The virtual characters of G form a ring.

PROOF. By part (i) of Proposition 8, the product of two characters is again a character, and the result easily follows from this observation. ∎

A *class function* on G is a function from G to $\mathbb{C}$ whose value within any conjugacy class is constant. For example, characters of $\mathbb{C}G$-modules are class functions. The set of class functions on G forms a $\mathbb{C}$-vector space of dimension r, where r is the number of conjugacy classes of G; an obvious basis for this vector space is the set of functions that attain the value 1 on a single conjugacy class and 0 on all other classes.

PROPOSITION 10. The irreducible characters of G form a basis for the space of class functions on G.

PROOF. By Proposition 5, the irreducible characters of G are linearly independent elements of the space of class functions; but their number is equal by Theorem 3 to the number of conjugacy classes of G, which is equal to the dimension of the space of class functions. ∎

If α and β are class functions on G, then their *inner product* is the complex number

$$(\alpha, \beta) = \frac{1}{|G|} \sum_{g \in G} \alpha(g)\overline{\beta(g)}.$$

This function (,) is, as the name suggests, an inner product on the space of class functions. That is, we have:

- $(\alpha, \alpha) \geq 0$ for all α, and $(\alpha, \alpha) = 0$ iff $\alpha = 0$.
- $(\alpha, \beta) = \overline{(\beta, \alpha)}$ for all α, β.
- $(\lambda\alpha, \beta) = \lambda(\alpha, \beta)$ for all α, β and all $\lambda \in \mathbb{C}$.
- $(\alpha_1 + \alpha_2, \beta) = (\alpha_1, \beta) + (\alpha_2, \beta)$ for all $\alpha_1, \alpha_2, \beta$.

The following are easy consequences of the above properties:

- $(\alpha, \lambda\beta) = \overline{\lambda}(\alpha, \beta)$ for all α, β and all $\lambda \in \mathbb{C}$.
- $(\alpha, \beta_1 + \beta_2) = (\alpha, \beta_1) + (\alpha, \beta_2)$ for all α, β_1, β_2.

We conclude this section with a result that gives some meaning to the inner product of two characters. We first require some notation and a lemma. If U is a $\mathbb{C}G$-module, then the set of elements of U on which G acts trivially is a $\mathbb{C}G$-submodule of U; we call this submodule U^G, and so $U^G = \{u \in U \mid gu = u \text{ for all } g \in G\}$.

LEMMA 11. If U is a $\mathbb{C}G$-module, then $\dim_{\mathbb{C}} U^G = \frac{1}{|G|} \sum_{g \in G} \chi_U(g)$.

PROOF. Let $a = \frac{1}{|G|} \sum_{g \in G} g \in \mathbb{C}G$. We have $ga = a$ for any $g \in G$, and from this we see that $a^2 = a$. If T is the linear transformation of U defined by a, then T must satisfy the equation $X^2 - X = 0$; consequently T is diagonalizable, and the only eigenvalues of T are 0 and 1. Let $U_1 \subseteq U$ be the eigenspace of T corresponding to the eigenvalue 1. If $u \in U_1$, then we have $gu = gau = au = u$ for any $g \in G$, and therefore $u \in U^G$. Conversely, suppose that $u \in U^G$. Then we see that $|G|au = (\sum_{g \in G} g)u = \sum_g gu = \sum_g u = |G|u$, and hence that $au = u$, which gives $u \in U_1$. Therefore $U^G = U_1$. However, the trace of T is clearly equal to the dimension of U_1, and the result now follows from the linearity of the trace map. ∎

THEOREM 12. We have $(\chi_U, \chi_V) = \dim_{\mathbb{C}} \mathrm{Hom}_{\mathbb{C}G}(U, V)$ for any $\mathbb{C}G$-modules U and V.

PROOF. We first observe that $\mathrm{Hom}_{\mathbb{C}G}(U, V)$ is a subspace of the $\mathbb{C}G$-module $\mathrm{Hom}(U, V)$. If $\varphi \in \mathrm{Hom}_{\mathbb{C}G}(U, V)$ and $g \in G$, then $(g\varphi)(u) = g\varphi(g^{-1}u) = gg^{-1}\varphi(u) = \varphi(u)$ for any $u \in U$; hence we have $g\varphi = \varphi$ for all $g \in G$, which shows that $\varphi \in \mathrm{Hom}(U, V)^G$. By reversing the argument we conclude that $\mathrm{Hom}_{\mathbb{C}G}(U, V) = \mathrm{Hom}(U, V)^G$. Therefore

$$\dim_{\mathbb{C}} \mathrm{Hom}_{\mathbb{C}G}(U, V) = \dim_{\mathbb{C}} \mathrm{Hom}(U, V)^G = \frac{1}{|G|} \sum_{g \in G} \chi_{\mathrm{Hom}(U,V)}(g)$$
$$= \frac{1}{|G|} \sum_{g \in G} \overline{\chi_U(g)}\chi_V(g)$$
$$= (\chi_V, \chi_U)$$

by Lemma 11 and part (iii) of Proposition 8. This implies that $(\chi_U, \chi_V) = \overline{(\chi_V, \chi_U)} = (\chi_V, \chi_U) = \dim_{\mathbb{C}} \mathrm{Hom}_{\mathbb{C}G}(U, V)$ since we now know that (χ_V, χ_U) is real-valued. ∎

EXERCISES

Throughout these exercises, G denotes a finite group, and all modules are finitely generated.

1. Let A be a semisimple $\mathbb{C}$-algebra. Let $[A, A]$ be the subspace of A spanned by the set $\{ab - ba \mid a, b \in A\}$. Show that the codimension of $[A, A]$ in A (which is just the dimension of $A/[A, A]$) is equal to the number of isomorphism classes of simple A-modules.
2. (cont.) Let $A = \mathbb{C}G$, and let $g_1, \ldots, g_r$ be representatives of the r conjugacy classes of G. Show that $\{g_i + [A, A] \mid 1 \le i \le r\}$ is a basis of $A/[A, A]$.
3. Show that if S is a simple $\mathbb{C}G$-module and U is a one-dimensional $\mathbb{C}G$-module, then $S \otimes U$ is simple.
4. Show that the set of isomorphism classes of the one-dimensional $\mathbb{C}G$-modules forms a group under the taking of tensor products.
5. Let χ be an irreducible character of G. Show that if λ is any $|G|$th root of unity, then $\{x \in G \mid \chi(x) = \lambda\chi(1)\} \unlhd G$.
6. Show that a $\mathbb{C}G$-module U is simple iff its dual U^* is simple. Conclude that the complex conjugate of an irreducible character is again an irreducible character.

If F is any field and U is any FG-module, then we define the character of U to be an F-valued function on U whose value at $g \in G$ is the trace of the F-endomorphism of U induced by g, exactly as was done in this section in the case $F = \mathbb{C}$.

7. Suppose that F has characteristic $p > 0$. Give an example to demonstrate that two FG-modules that have the same character need not be isomorphic.

8. Suppose that F has characteristic $p > 0$. Show that if U is an FG-module then we have $\chi_U(g) = \chi_U(g_{p'})$ for any $g \in G$, where $g_{p'}$ is the p'-part of g as defined in Exercise 1.3.

15. The Character Table

We established in Section 14 that any character of G is a $\mathbb{Z}$-linear combination of the r irreducible characters $\chi_1, \ldots, \chi_r$ of G, where r is by Theorem 14.3 equal to the number of conjugacy classes of G. Since each irreducible character is specified by its value on each conjugacy class of G, it follows that the characters of G are completely determined by an $r \times r$ array giving the values of the r irreducible characters on the r conjugacy classes of G. This array is called the *character table* of G. Of course, the character table of G is well-defined only up to reorderings of the rows and columns.

If $\mathcal{X}$ is the character table of G, then $\mathcal{X} = (\chi_i(g_j))_{1 \le i,j \le r}$, where $g_1, \ldots, g_r$ are representatives of the r conjugacy classes of G. By convention, we always set $g_1 = 1$, so that the first column of the character table consists of the degrees of G. We will generally write $\mathcal{X}$ in the form

$$
\begin{array}{c|cccc}
 & 1 & k_2 & \cdots & k_r \\
 & 1 & g_2 & \cdots & g_r \\
\hline
\chi_1 & 1 & 1 & \cdots & 1 \\
\chi_2 & f_2 & \chi_2(g_2) & \cdots & \chi_2(g_r) \\
\vdots & \vdots & \vdots & \ddots & \vdots \\
\chi_r & f_r & \chi_r(g_2) & \cdots & \chi_r(g_r)
\end{array}
$$

where the f_i are the degrees of G, and $k_i = |G : C_G(g_i)|$ is (by Proposition 3.11) the order of the conjugacy class of g_i for each i.

(Recall that we always take χ_1 to be the principal character.)

We now continue to establish some fundamental properties of characters.

ROW ORTHOGONALITY THEOREM. $(\chi_i, \chi_j) = \delta_{ij}$ for any i and j.

PROOF. Let $S_1, \ldots, S_r$ be the distinct simple $\mathbb{C}G$-modules. By Theorem 14.12, we have $(\chi_i, \chi_j) = \dim_{\mathbb{C}} \text{Hom}_{\mathbb{C}G}(S_i, S_j)$ for any i and j. For each i, we have $\text{Hom}_{\mathbb{C}G}(S_i, S_i) = \text{End}_{\mathbb{C}G}(S_i) \cong \mathbb{C}$ by Lemma 13.14, and if $i \neq j$ then $\text{Hom}_{\mathbb{C}G}(S_i, S_j) = 0$ by Schur's lemma. ■

In other words, this theorem asserts that

$$\delta_{ij} = \frac{1}{|G|} \sum_{g \in G} \chi_i(g)\overline{\chi_j(g)} = \frac{1}{|G|} \sum_{t=1}^{r} k_t \chi_i(g_t)\overline{\chi_j(g_t)}$$

for any i and j, where the g_t are conjugacy class representatives and the k_t are the orders of the conjugacy classes. We can interpret this as saying that the rows of the character table are, when considered as vectors in $\mathbb{C}^r$, orthogonal with respect to an inner product that differs slightly from the standard one, and it is this interpretation that lends its name to the above result.

Row orthogonality has a number of important consequences:

COROLLARY 1. The irreducible characters of G form an orthonormal basis for the vector space of class functions on G.

PROOF. This follows immediately from Proposition 14.10 and the row orthogonality theorem. ■

COROLLARY 2. If $\alpha = \sum_i a_i \chi_i$ and $\beta = \sum_j b_j \chi_j$ are virtual characters of G, then $(\alpha, \beta) = \sum_i a_i b_i$.

PROOF. We have

$$(\alpha, \beta) = \left(\sum_{i=1}^{r} a_i \chi_i, \sum_{j=1}^{r} b_i \chi_j \right) = \sum_{i=1}^{r} \sum_{j=1}^{r} a_i b_j (\chi_i, \chi_j) = \sum_{i=1}^{r} \sum_{j=1}^{r} a_i b_j \delta_{ij}$$
$$= \sum_{i=1}^{r} a_i b_i$$

by row orthogonality. ■

COROLLARY 3. If α is a character of G and $n \in \{1, 2, 3\}$, then $(\alpha, \alpha) = n$ iff α is a sum of n irreducible characters.

PROOF. Write $\alpha = \sum_{i=1}^{r} a_i \chi_i$, where the a_i are non-negative integers. Since $(\alpha, \alpha) = \sum_i a_i^2$ by Corollary 2, we see that if $(\alpha, \alpha) = n$, then we must have $a_j = 1$ for exactly n numbers $1 \le j \le r$, and $a_i = 0$ for all other i, in which case α is a sum of n irreducible characters. The converse follows directly from Corollary 2. ∎

COROLLARY 4. If α is a virtual character of G, then each χ_j appears with coefficient (α, χ_j) in the unique expression of α as a $\mathbb{Z}$-linear combination of the irreducible characters of G.

PROOF. Write $\alpha = \sum_{i=1}^{r} a_i \chi_i$, where the a_i are integers; then $(\alpha, \chi_j) = a_j$ by Corollary 2. ∎

PROPOSITION 5. If α is a linear character of G and χ is an irreducible character of G, then $\alpha\chi$ is an irreducible character of G.

(This result also follows from Exercise 14.3 and part (i) of Proposition 14.8, but here we provide an alternate proof.)

PROOF. Since α is linear, it follows from part (iii) of Proposition 14.4 that $\alpha(g)$ is a root of unity for any $g \in G$, and in particular that $1 = |\alpha(g)| = \alpha(g)\overline{\alpha(g)}$ for every $g \in G$. We now have

$$(\alpha\chi, \alpha\chi) = \frac{1}{|G|} \sum_{g \in G} \alpha(g)\chi(g)\overline{\alpha(g)\chi(g)}$$

$$= \frac{1}{|G|} \sum_{g \in G} \chi(g)\overline{\chi(g)}\alpha(g)\overline{\alpha(g)}$$

$$= \frac{1}{|G|} \sum_{g \in G} \chi(g)\overline{\chi(g)} = (\chi, \chi) = 1$$

by row orthogonality, and hence $\alpha\chi$ is irreducible by Corollary 3. ∎

The following theorem implies that the columns of the character table are orthogonal when considered as vectors in $\mathbb{C}^r$ under the standard inner product.

COLUMN ORTHOGONALITY THEOREM. If $g_1, \ldots, g_r$ are a set of conjugacy class representatives of G, and $k_1, \ldots, k_r$ are the orders of the conjugacy classes, then for any $1 \le i, j \le r$ we have

$$\sum_{t=1}^{r} \chi_t(g_i)\overline{\chi_t(g_j)} = \frac{|G|}{k_i}\delta_{ij} = |C_G(g_i)|\delta_{ij}.$$

PROOF. Let $\mathcal{X} = (\chi_i(g_j))_{1 \le i,j \le r}$ be the character table of G, and let K be the $r \times r$ diagonal matrix having $(k_1, \ldots, k_r)$ as its main diagonal. Then $(\mathcal{X}K)_{ij} = \sum_{\ell=1}^{r} \chi_i(g_\ell)(K)_{\ell j} = \chi_i(g_j)k_j$ for any i and j, and thus we have

$$(\mathcal{X}K\overline{\mathcal{X}}^t)_{ij} = \sum_{\ell=1}^{r} \chi_i(g_\ell)k_\ell(\overline{\mathcal{X}}^t)_{\ell j} = \sum_{\ell=1}^{r} k_\ell \chi_i(g_\ell)\overline{\chi_j(g_\ell)}$$
$$= \sum_{g \in G} \chi_i(g)\overline{\chi_j(g)}$$
$$= |G|(\chi_i, \chi_j) = |G|\delta_{ij}$$

for any i and j by row orthogonality. Therefore $\mathcal{X}K\overline{\mathcal{X}}^t = |G|I$, where I is the identity matrix. We leave it to the reader to verify that if A and B are matrices such that AB is a non-zero scalar matrix, then $BA = AB$. Thus $K\overline{\mathcal{X}}^t\mathcal{X} = |G|I$, and hence for any i and j we have $|G|\delta_{ji} = \sum_{\ell=1}^{r}(K\overline{\mathcal{X}}^t)_{j\ell}\mathcal{X}_{\ell i} = \sum_{\ell=1}^{r} k_j \overline{\chi_\ell(g_j)}\chi_\ell(g_i)$ as required. ∎

As we shall see later in this section, the orthogonality relations often make it possible to construct a character table even if little is known about the group in question. This raises the following question, which we now address: What information about a group can be obtained from its character table? We first require some information about the connection between the representation theory of a group and that of its quotient groups.

LEMMA 6. Let $N \trianglelefteq G$, and let U be a $\mathbb{C}(G/N)$-module. Then U admits a canonical $\mathbb{C}G$-module structure, with a subspace of U being a $\mathbb{C}G$-submodule iff it is a $\mathbb{C}(G/N)$-submodule. If ψ is the character of the $\mathbb{C}(G/N)$-module U, then the character of the $\mathbb{C}G$-module U is $\psi \circ \eta$, where $\eta \colon G \to G/N$ is the natural map.

PROOF. Given $g \in G$ and $u \in U$, we define $gu = (gN)u$; this gives U a $\mathbb{C}G$-module structure in which the $\mathbb{C}G$-submodules of U are exactly the $\mathbb{C}(G/N)$-submodules of U. If $g \in G$, then the linear transformation of U induced by g under the action of G on U is exactly the same as that induced by $\eta(g) = gN$ under the action of G/N on U, which implies the statement concerning characters. ∎

We define $K_\chi = \{x \in G \mid \chi(x) = \chi(1)\}$ for any character χ of G; we have $K_\chi \trianglelefteq G$ by part (vi) of Proposition 14.4. We call K_χ the *kernel* of χ, as it is the kernel of the corresponding representation. We write K_i instead of K_{χ_i}.

PROPOSITION 7. The normal subgroups of G are exactly the sets of the form $\bigcap_{i \in I} K_i$ for some $I \subseteq \{1, \dots, r\}$.

PROOF. Let $N \trianglelefteq G$, and let $U = \mathbb{C}(G/N)$; let ψ be the character of U considered as a $\mathbb{C}(G/N)$-module, and let χ be the character of U considered via Lemma 6 as a $\mathbb{C}G$-module. Since ψ is the regular character of G/N, it follows from Lemma 6 that $\chi(g) = \chi(1)$ iff $g \in N$; therefore $K_\chi = N$. Write $\chi = \sum_i a_i \chi_i$ for some non-negative integers a_i. We observe via part (v) of Proposition 14.4 that we have $|\chi(g)| \leq \sum_i a_i |\chi_i(g)| \leq \sum_i a_i \chi_i(1) = \chi(1)$ for any $g \in G$. It follows from these inequalities and again from part (v) of Proposition 14.4 that $g \in K_\chi$ iff $g \in K_i$ for every i for which $a_i > 0$. Therefore $N = \bigcap_{i \in I} K_i$, where $I = \{1 \leq i \leq r \mid a_i > 0\}$. Conversely, since each K_i is normal in G, it follows from Proposition 1.7 that $\bigcap_{i \in I} K_i \trianglelefteq G$ for any $I \subseteq \{1, \dots, r\}$. ∎

COROLLARY 8. G is simple iff the only irreducible character χ_i for which $\chi_i(g) = \chi_i(1)$ for some $1 \neq g \in G$ is the principal character χ_1.

PROOF. If G is simple and $\chi_i(g) = \chi_i(1)$ for some $i > 1$ and some $1 \neq g \in G$, then as $g \in K_i \trianglelefteq G$ we obtain a contradiction. Conversely, if G is not simple, then there is some $1 \neq g \in G$ lying in some non-trivial proper normal subgroup N; by Proposition 7 we must have $N \leqslant K_i$ for some $i > 1$, in which case $\chi_i(g) = \chi_i(1)$. ∎

COROLLARY 9. The character table of G can be used to determine whether or not G is solvable.

PROOF. It follows from Proposition 7 that the character table of G enables us to determine all of the normal subgroups of G and all of the inclusion relations between the normal subgroups. Hence we can determine all normal series of G and the orders of the terms thereof. In particular, we can determine whether or not G has a normal series whose successive quotients are p-groups, which by Corollary 11.7 is a criterion for solvability. ∎

We define $Z_\chi = \{x \in G \mid |\chi(x)| = \chi(1)\}$ for any character χ of G. Again we write Z_i instead of Z_{χ_i}.

LEMMA 10. $Z_\chi \leqslant G$ for any character χ of G, and if in addition χ is irreducible, then $Z_\chi/K_\chi = Z(G/K_\chi)$.

PROOF. Let $g \in G$. We see from part (iv) of Proposition 14.4 that if $g \in Z_\chi$, then $g^{-1} \in Z_\chi$. Since $\chi(g)$ is a sum of $\chi(1)$ roots of unity by part (iii) of Proposition 14.4, we see that $|\chi(g)| = \chi(1)$ iff g has exactly one eigenvalue. If $g \in Z_\chi$, let this eigenvalue be $\lambda(g)$, so that if U is the $\mathbb{C}G$-module corresponding to χ, then we have $gu = \lambda(g)u$ for all $u \in U$. We now see that if $g, h \in Z_\chi$, then $(gh)u = \lambda(g)\lambda(h)u$ for all $u \in U$; hence $\chi(gh) = \chi(1)\lambda(g)\lambda(h)$, and thus $|\chi(gh)| = \chi(1)$, giving $gh \in Z_\chi$. Therefore $Z_\chi \leqslant G$. Now if $\rho: G \to \mathrm{GL}(U)$ is the representation corresponding to χ, then for any $g \in Z_\chi$, the matrix of $\rho(g)$ (with respect to any $\mathbb{C}$-basis of U) will be scalar, and hence $\rho(g) \in Z(\rho(G))$. Since $\rho(G) \cong G/K_\chi$, it follows that $Z_\chi/K_\chi \leqslant Z(G/K_\chi)$.

Now suppose that χ is irreducible. If $gK_\chi \in Z(G/K_\chi)$, then $\rho(g)$ commutes with $\rho(x)$ for every $x \in G$, and consequently the map sending $u \in U$ to gu is a $\mathbb{C}G$-endomorphism of U. But U is simple, so we have $\mathrm{End}_{\mathbb{C}G}(U) \cong \mathbb{C}$ by Schur's lemma. Therefore, there is some complex root of unity μ such that $gu = \mu u$ for all $u \in U$. We now have $\chi(g) = \chi(1)\mu$, which gives $|\chi(g)| = \chi(1)$ and hence $g \in Z_\chi$. Therefore $Z_\chi/K_\chi = Z(G/K_\chi)$. ∎

COROLLARY 11. If G is non-abelian and simple, then $Z_i = 1$ for all $i > 1$.

PROOF. If $i > 1$ then $K_i = 1$, so $Z_i = Z(G) = 1$ by Lemma 10. ∎

PROPOSITION 12. $Z(G) = \bigcap\limits_{i=1}^{r} Z_i$.

PROOF. If χ is any character of G, then $Z(G)K_\chi/K_\chi \leqslant Z(G/K_\chi)$. Thus by Lemma 10 we have $Z(G)K_i/K_i \leqslant Z_i/K_i$, and consequently $Z(G) \leqslant Z_i$, for every i. Conversely, suppose that $g \in Z_i$ for all i. Since $Z_i/K_i = Z(G/K_i)$ by Lemma 10, it follows that for any $x \in G$ we have $[g, x] \in K_i$ for all i. But it follows from Proposition 7 that $K_1 \cap \ldots \cap K_r = 1$; therefore $[g, x] = 1$ for all $x \in G$, and hence $g \in Z(G)$. ∎

LEMMA 13. If $N \trianglelefteq G$, then the irreducible characters of G/N can be determined from those of G.

PROOF. Let χ be any irreducible character of G whose kernel contains N, and let U be the $\mathbb{C}G$-module corresponding to χ. Then since N acts on U via the identity, we can give U a $\mathbb{C}(G/N)$-module structure by $(gN)u = gu$ for $gN \in G/N$ and $u \in U$. As U is a simple $\mathbb{C}G$-module, it follows that U is also simple as a $\mathbb{C}(G/N)$-module, and the character of the $\mathbb{C}(G/N)$-module U sends gN to $\chi(g)$. Therefore, those irreducible characters of G whose kernel contains N give rise to irreducible characters of G/N. But by Lemma 6, every irreducible character of G/N gives rise to an irreducible character of G whose kernel contains N.

It now follows that all irreducible characters of G/N can be determined from those of G, in the following sense: What we can determine is the number of irreducible characters of G/N, and their values on any element of G/N. We cannot readily determine the actual conjugacy classes of G/N, or their orders, although it is certainly true that conjugate elements of G have images in G/N that are conjugate, and hence the image in G/N of a class of G is a union of classes of G/N. Nonetheless, the information that can be gained from the character table of G about the characters of G/N is sufficient, for instance, to determine $Z(G/N)$ via Proposition 12. ∎

COROLLARY 14. The character table of G can be used to determine whether or not G is nilpotent.

PROOF. Consider the upper central series $1 \leqslant \mathcal{Z}_1 \leqslant \mathcal{Z}_2 \leqslant \ldots$ of G defined in the further exercises to Section 11. Since we have $\mathcal{Z}_i \trianglelefteq G$ and $\mathcal{Z}_i/\mathcal{Z}_{i-1} = Z(G/\mathcal{Z}_{i-1})$ for each i, we see from Proposition 12 that each $\mathcal{Z}_i$ can be determined from the irreducible characters of $G/\mathcal{Z}_{i-1}$, which by Lemma 13 can be determined from the character table of G. Consequently, we can use the character table of G to determine every term $\mathcal{Z}_i$ of the upper central series, and the result follows by Exercise 11.10. (This argument also shows that if G is nilpotent, then the character table of G allows us to determine the nilpotency class of G.) ∎

Observe that in Corollary 9 we assumed that we knew both the irreducible characters of G and the orders of the conjugacy classes, whereas in Corollary 14 we did not need to know the orders of the classes. As evidenced by Lemma 13, there may be circumstances in which we can construct a character table without knowing the orders of the conjugacy classes. There is a theorem of G. Higman (see [15,

p. 136]) which asserts that if we know the irreducible characters of G, but not the orders of the conjugacy classes, then we can at least determine the sets of prime divisors of the orders of the conjugacy classes.

The remainder of this section consists of a series of examples in which we develop methods of finding characters and use these methods to construct the character tables of various groups.

EXAMPLE 1. Let $G = <g> \cong \mathbf{Z_n}$ for some $n \in \mathbb{N}$, and let λ be a primitive nth root of unity. For each $1 \le i \le n$, let V_i be a one-dimensional $\mathbb{C}$-vector space, and let g act on V_i by multiplication by λ^{i-1}; since G is cyclic, this definition completely determines a $\mathbb{C}G$-module structure on V_i. Each V_i, being one-dimensional, is a simple $\mathbb{C}G$-module. If χ_i is the character of V_i, then $\chi_i(g) = \lambda^{i-1}$, and hence $\chi_i(g^a) = \lambda^{a(i-1)}$ for any a. The characters $\chi_1,\ldots,\chi_n$ are n distinct linear characters of G, and since G can have at most $|G| = n$ irreducible characters, we see that the χ_i are precisely the irreducible characters of G. Thus, the character table of G is

	1	1	1	$\ldots$	1
	1	g	g^2	$\ldots$	g^{n-1}
χ_1	1	1	1	$\ldots$	1
χ_2	1	λ	λ^2	$\ldots$	λ^{n-1}
χ_3	1	λ^2	λ^4	$\ldots$	λ^{n-2}
$\vdots$	$\vdots$	$\vdots$	$\vdots$	$\ddots$	$\vdots$
χ_n	1	λ^{n-1}	λ^{n-2}	$\ldots$	λ

Observe that the set $\{\chi_1,\ldots,\chi_n\}$ is a cyclic group under multiplication of characters, with generator χ_2; we have $\chi_i = \chi_2^{i-1}$ for any i.

EXAMPLE 2. Let G and H be groups with respective irreducible characters $\chi_1,\ldots,\chi_r$ and $\psi_1,\ldots,\psi_s$. We wish to determine the irreducible characters of $G \times H$. We see that two elements (x,y) and (x',y') of $G \times H$ are conjugate iff x and x' are conjugate in G and y and y' are conjugate in H. Since by Theorem 14.3, G and H have r and s conjugacy classes, respectively, we conclude that $G \times H$ has rs conjugacy classes, with each class of $G \times H$ being the product of a class of G and a class of H; hence by Theorem 14.3, $G \times H$ has rs irreducible characters.

Let $S_1, \ldots, S_r$ and $T_1, \ldots, T_s$ be the distinct simple $\mathbb{C}G$- and $\mathbb{C}H$-modules, respectively. For each i and j, we give $S_i \otimes T_j$ the structure of a $\mathbb{C}(G \times H)$-module by $(g, h)(s \otimes t) = gs \otimes ht$ and linear extension. Let τ_{ij} be the character of $S_i \otimes T_j$ for each i and j. By imitating the proof of part (i) of Proposition 14.8, we find that $\tau_{ij}(g, h) = \chi_i(g)\psi_j(h)$; we adopt the notation $\tau_{ij} = \chi_i \times \psi_j$. We note that the τ_{ij} are distinct and that we can recover the χ_i and ψ_j from the τ_{ij} by appropriate restriction. Now for any i, i', j, j' we have

$$(\tau_{ij}, \tau_{i'j'}) = \frac{1}{|G \times H|} \sum_{(g,h) \in G \times H} \tau_{ij}(g, h)\overline{\tau_{i'j'}(g, h)}$$

$$= \frac{1}{|G||H|} \sum_{g \in G} \sum_{h \in H} \chi_i(g)\psi_j(h)\overline{\chi_{i'}(g)}\overline{\varphi_{j'}(h)}$$

$$= \left(\frac{1}{|G|} \sum_{g \in G} \chi_i(g)\overline{\chi_{i'}(g)}\right)\left(\frac{1}{|H|} \sum_{h \in H} \psi_j(h)\overline{\psi_{j'}(h)}\right)$$

$$= (\chi_i, \chi_{i'})(\psi_j, \psi_{j'}) = \delta_{ii'}\delta_{jj'}$$

by row orthogonality. In particular, we have $(\tau_{ij}, \tau_{ij}) = 1$ for each i and j, which by Corollary 3 implies that τ_{ij} is an irreducible character. Therefore the τ_{ij} are the rs irreducible characters of $G \times H$; that is, the set of irreducible characters of $G \times H$ is $\{\chi_i \times \psi_j\}_{i,j}$.

EXAMPLE 3. Suppose that G is abelian. Then every conjugacy class of G contains exactly one element, and consequently G has $|G|$ irreducible characters by Theorem 14.3. But $\sum_{i=1}^{|G|} f_i^2 = |G|$ by Corollary 14.2, so we must have $f_i = 1$ for each i. Therefore all irreducible characters of G are linear. By Corollary 8.8, G is a direct product of cyclic p-groups; let these cyclic p-groups be of orders $p_1^{a_1}, \ldots, p_t^{a_t}$ with respective generators $g_1, \ldots, g_t$. We could determine the character table of G from those of the cyclic p-groups using Examples 1 and 2, but there is an alternate method, which we now describe. A linear character χ of G is just a homomorphism from G to $\mathbb{C}^\times$, and hence to define such a χ it suffices to specify $\chi(g_i)$ for each i, with the only restriction on $\chi(g_i)$ being that it is a $p_i^{a_i}$-th root of unity. Therefore, there is a bijective correspondence between irreducible characters of G and ordered t-tuples $(\lambda_1, \ldots, \lambda_t)$ in which each λ_i is a $p_i^{a_i}$-th root of unity.

As a simple example, let $G = <a, b> \cong \mathbf{Z_2} \times \mathbf{Z_2}$. Both a and b have order 2, so by the above paragraph we see that the four irreducible characters of G correspond to the ordered pairs $(1,1), (1,-1), (-1,1)$, and $(-1,-1)$. Therefore, the character table for G is

$$
\begin{array}{c|cccc}
 & 1 & 1 & 1 & 1 \\
 & 1 & a & b & ab \\
\hline
\chi_1 & 1 & 1 & 1 & 1 \\
\chi_2 & 1 & 1 & -1 & -1 \\
\chi_3 & 1 & -1 & 1 & -1 \\
\chi_4 & 1 & -1 & -1 & 1
\end{array}
$$

Here the second and third columns correspond to the ordered pairs given above, and the fourth column is wholly determined by the previous two columns.

EXAMPLE 4. Let X be a finite G-set. We saw in Section 12 that the vector space $\mathbb{C}X$ having basis X is a $\mathbb{C}G$-module. We wish to determine the character π of $\mathbb{C}X$. Consider the matrix, with respect to the basis X, of the linear transformation of $\mathbb{C}X$ defined by $g \in G$. Since $gx \in X$ for every $x \in X$, this matrix is a permutation matrix; but $\pi(g)$ is the sum of the diagonal entries of this matrix, so we conclude that $\pi(g)$ equals the number of elements of X which are fixed by g.

Let χ_1 be the principal character of G. We have

$$
(\pi, \chi_1) = \frac{1}{|G|} \sum_{g \in G} \pi(g)\chi_1(g) = \frac{1}{|G|} \sum_{g \in G} \pi(g)
$$

$$
= \frac{1}{|G|} \sum_{g \in G} |\{x \in X \mid gx = x\}|
$$

$$
= \frac{1}{|G|} |\{(g,x) \in G \times X \mid gx = x\}|
$$

$$
= \frac{1}{|G|} \sum_{x \in X} |\{g \in G \mid gx = x\}|.
$$

For each $x \in X$, let G_x be the stabilizer of x; using Corollary 3.5, we now have

$$
(\pi, \chi_1) = \sum_{x \in X} \frac{|G_x|}{|G|} = \sum_{\text{orbits } \mathcal{O}} \sum_{x \in \mathcal{O}} \frac{1}{|G : G_x|}
$$

$$
= \sum_{\text{orbits } \mathcal{O}} \sum_{x \in \mathcal{O}} \frac{1}{|\mathcal{O}|} = \sum_{\text{orbits } \mathcal{O}} |\mathcal{O}| \cdot \frac{1}{|\mathcal{O}|} = \sum_{\text{orbits } \mathcal{O}} 1,
$$

which allows us to conclude that (π, χ_1) is equal to the number of orbits of X under the action of G. (This result is often attributed to Burnside, although as argued in [21] it would be more accurate to call this result the Cauchy-Frobenius lemma.) In particular, if G acts transitively on X, then we see via Corollary 4 that the unique expression of the character $\pi - \chi_1$ as a linear combination of the irreducible characters of G does not involve χ_1.

EXAMPLE 5. The group G/G' is abelian by Proposition 2.6, and hence (as seen in Example 3) all of its irreducible characters are linear. It follows from Lemma 6 that each linear character of G/G' can be lifted to a linear character of G, and that the lifts of distinct characters are distinct. Now let χ be a linear character of G, so that $\chi \colon G \to \mathbb{C}^\times$ is a homomorphism whose kernel is K_χ. Then G/K_χ is abelian, being isomorphic via χ with a subgroup of $\mathbb{C}^\times$, and hence $G' \leqslant K_\chi$ by Proposition 2.6. We can now define a linear character ψ of G/G' whose lift is χ by $\psi(gG') = \chi(g)$. We conclude that every linear character of G is the lift of a linear character of G/G'.

EXAMPLE 6. Consider Σ_3. The conjugacy classes of Σ_3 are $\{1\}$, $\{(1\ 2), (1\ 3), (2\ 3)\}$, and $\{(1\ 2\ 3), (1\ 3\ 2)\}$. (This follows from Proposition 1.10.) We have $\Sigma_3/A_3 \cong \mathbb{Z}_2$, so Σ_3 has two linear characters, which are lifted from the characters of $\mathbb{Z}_2$ as in Lemma 6. Letting these characters be χ_1 and χ_2, we have

	1	3	2
	1	(1 2)	(1 2 3)
χ_1	1	1	1
χ_2	1	-1	1
χ_3			

Since $6 = |\Sigma_3| = f_1^2 + f_2^2 + f_3^2 = 2 + f_3^2$ by Corollary 14.2, we have $f_3 = 2$. Column orthogonality now gives

$$0 = \sum_{i=1}^{3} f_i \chi_i((1\ 2)) = 1 \cdot 1 + 1 \cdot (-1) + 2\chi_3((1\ 2))$$

and

$$0 = \sum_{i=1}^{3} f_i \chi_i((1\ 2\ 3)) = 1 \cdot 1 + 1 \cdot 1 + 2\chi_3((1\ 2)),$$

from which we see that the character table of Σ_3 is

	1	3	2
	1	(1 2)	(1 2 3)
χ_1	1	1	1
χ_2	1	-1	1
χ_3	2	0	-1

EXAMPLE 7. Consider Σ_4. Recall that Σ_4 has a normal subgroup of order 4, namely $K = \{1, (1\ 2)(3\ 4), (1\ 3)(2\ 4), (1\ 4)(2\ 3)\}$. The subgroup Σ_3 of Σ_4 intersects K trivially, and $|\Sigma_4| = |\Sigma_3||K|$; therefore $\Sigma_4 = K \rtimes \Sigma_3$. In particular, we have $\Sigma_4/K \cong \Sigma_3$, and thus we can obtain irreducible characters of Σ_4 from the irreducible characters of Σ_3 as in Lemma 6.

By Proposition 1.10, each conjugacy class of Σ_4 consists of all elements having a given cycle structure. Hence Σ_4 has 5 classes, with representatives 1, (1 2)(3 4), (1 2 3), (1 2), and (1 2 3 4); the orders of the classes are 1, 3, 8, 6, and 6, respectively. Let χ_1, χ_2, and χ_3 be the irreducible characters of Σ_4 obtained from those of Σ_3. To determine these characters, we need information about the images in Σ_4/K of the class representatives. We easily see that the images of 1 and (1 2)(3 4) are trivial, that the image of (1 2 3) has order 3, and that the images of (1 2) and (1 2 3 4) have order 2. Using this observation and the character table for Σ_3 obtained in Example 6, we obtain the following partial character table for Σ_4:

	1	3	8	6	6
	1	(1 2)(3 4)	(1 2 3)	(1 2)	(1 2 3 4)
χ_1	1	1	1	1	1
χ_2	1	1	1	-1	-1
χ_3	2	2	-1	0	0
χ_4					
χ_5					

Now Σ_4 acts transitively on the set $X = \{1, 2, 3, 4\}$; let π be the character of the $\mathbb{C}G$-module $\mathbb{C}X$. Since $\pi(g)$ equals the number of fixed points under the action of g on X by Example 4, we have

	1	3	8	6	6
	1	(1 2)(3 4)	(1 2 3)	(1 2)	(1 2 3 4)
π	4	0	1	2	0
$\pi - \chi_1$	3	-1	0	1	-1

We also have

$$(\pi - \chi_1, \pi - \chi_1) = \frac{1}{|\Sigma_4|} \sum_{i=1}^{5} k_i[(\pi - \chi_1)(g_i)]^2$$

$$= \frac{1}{24}(1 \cdot 9 + 3 \cdot 1 + 8 \cdot 0 + 6 \cdot 1 + 6 \cdot 1) = 1,$$

and hence $\pi - \chi_1$ is irreducible by Corollary 3. Let $\chi_4 = \pi - \chi_1$. Now $\chi_2\chi_4 \neq \chi_4$, and $\chi_2\chi_4$ is irreducible by Proposition 5; therefore $\chi_5 = \chi_2\chi_4$, and hence the character table of Σ_4 is

	1	3	8	6	6
	1	(1 2)(3 4)	(1 2 3)	(1 2)	(1 2 3 4)
χ_1	1	1	1	1	1
χ_2	1	1	1	-1	-1
χ_3	2	2	-1	0	0
χ_4	3	-1	0	1	-1
χ_5	3	-1	0	-1	1

EXAMPLE 8. Let G be a non-abelian group of order 8. We will show that this information completely determines the character table of G. Since there are two isomorphism classes of non-abelian groups of order 8 (see page 79), this will imply that the character table does not specify a group up to isomorphism.

We have $|Z(G)| = 2$ by Theorem 8.1 and Lemma 8.2, and hence $Z(G) \cong \mathbf{Z_2}$. Since G is non-abelian and $G/Z(G)$ is abelian (being a group of order 4), this forces $G' = Z(G)$ by Proposition 2.6. Now $|G/G'| = 4$, and by Lemma 8.2 G/G' cannot be cyclic since G is non-abelian and $G' = Z(G)$, so we must have $G/G' \cong \mathbf{Z_2} \times \mathbf{Z_2}$. By Examples 3 and 5, we now know that G has exactly 4 linear characters. By Corollary 14.2, we have $\sum_{i=1}^{r} f_i^2 = 8$, where $f_i = 1$ for $1 \leq i \leq 4$ and $f_i > 1$ for $i > 4$; this forces $r = 5$ and $f_5 = 2$.

Let x be the generator of G'. As $G' = Z(G)$, we see that 1 and x are the only elements of G lying in single-element conjugacy classes, and hence each of the other three classes must have order 2. Let a,

b, and c be representatives of the multi-element classes. Since G/G' also has four conjugacy classes, we see that the images of a, b, and c in G/G' must lie in distinct conjugacy classes. Using the character table for $\mathbf{Z_2} \times \mathbf{Z_2}$ determined in Example 3, we can now obtain a partial character table for G by lifting:

	1	1	2	2	2
	1	x	a	b	c
χ_1	1	1	1	1	1
χ_2	1	1	1	-1	-1
χ_3	1	1	-1	1	-1
χ_4	1	1	-1	-1	1
χ_5	2				

We see via column orthogonality that $\chi_5(a) = \chi_5(b) = \chi_5(c) = 0$ and that $\chi_5(x) = -2$, completing the character table.

EXAMPLE 9. Consider the group A_5. Using Proposition 1.10, we find that the elements of A_5 form 4 conjugacy classes in Σ_5, with representatives 1, (1 2)(3 4), (1 2 3), and (1 2 3 4 5) and respective orders 1, 15, 20, and 24. However, elements of A_5 that are conjugate in Σ_5 may not be conjugate in A_5.

Let $x \in A_5$, and let $\mathcal{K}$ be the conjugacy class in Σ_5 of x. We see from Proposition 3.13 that $|\mathcal{K}| = |\Sigma_5 : C_{\Sigma_5}(x)|$ and that the conjugacy class in A_5 of x is contained in $\mathcal{K}$ and has order $|A_5 : C_{A_5}(x)|$. Suppose first that $C_{\Sigma_5}(x) \nleq A_5$. This forces $A_5 C_{\Sigma_5}(x) = \Sigma_5$ since A_5 is a maximal subgroup of Σ_5. The first isomorphism theorem gives $\Sigma_5/A_5 = A_5 C_{\Sigma_5}(x)/A_5 \cong C_{\Sigma_5}(x)/C_{\Sigma_5}(x) \cap A_5 = C_{\Sigma_5}(x)/C_{A_5}(x)$, and hence

$$|A_5 : C_{A_5}(x)| = \frac{|\Sigma_5 : C_{A_5}(x)|}{|\Sigma_5 : A_5|} = \frac{|\Sigma_5 : C_{A_5}(x)|}{|C_{\Sigma_5}(x) : C_{A_5}(x)|} = |\Sigma_5 : C_{\Sigma_5}(x)|$$
$$= |\mathcal{K}|.$$

Thus $\mathcal{K}$ is the conjugacy class in A_5 of x. Now suppose $C_{\Sigma_5}(x) \leqslant A_5$. Then we have $C_{A_5}(x) = C_{\Sigma_5}(x) \cap A_5 = C_{\Sigma_5}(x)$, and hence

$$|A_5 : C_{A_5}(x)| = |A_5 : C_{\Sigma_5}(x)| = \tfrac{1}{2}|\Sigma_5 : C_{\Sigma_5}(x)| = \tfrac{1}{2}|\mathcal{K}|,$$

from which we see that $\mathcal{K}$ splits into two conjugacy classes in A_5 of equal cardinality.

We easily see that each of the elements 1, (1 2)(3 4), and (1 2 3) commutes with some odd permutation in Σ_5; for example, (4 5) and

(1 2 3) commute. Hence by the above paragraph, the conjugacy classes in A_5 of these elements are the respective classes in Σ_5. Since $24 \nmid 60$, we can now conclude that the conjugacy class in Σ_5 of (1 2 3 4 5) splits into two classes in A_5, and hence that the conjugacy classes of A_5 have orders $1, 15, 20, 12$, and 12. (Using this fact and the remark made after Proposition 3.13, we could now give a slick proof of the simplicity of A_5.) We will now show that $x = (1\ 2\ 3\ 4\ 5)$ and $x^2 = (1\ 3\ 5\ 2\ 4)$ are not conjugate in A_5, thereby establishing that the conjugacy class in Σ_5 of x splits in A_5 into the class containing x and the class containing x^2.

Suppose that $g \in A_5$ is such that $gxg^{-1} = x^2$. Then we have $g<x>g^{-1} = <x^2> = <x>$, and hence $g \in N_{A_5}(<x>)$. Now $<x>$ is a Sylow 5-subgroup of A_5, and we observed in the proof of Theorem 7.4 that A_5 has 6 Sylow 5-subgroups. Thus $|N_{A_5}(<x>)| = 10$, and we find that $N_{A_5}(<x>) = <x> \rtimes <t> \cong D_{10}$, where $t = (2\ 5)(3\ 4)$. But $txt^{-1} = x^{-1} \neq x^2$; from this, we see that there is no element $g \in N_{A_5}(<x>)$ such that $gxg^{-1} = x^2$. Contradiction; therefore x and x^2 are not conjugate in A_5.

We now see that A_5 has 5 conjugacy classes, with representatives 1, (1 2)(3 4), (1 2 3), (1 2 3 4 5), and (1 3 5 2 4) and respective orders $1, 15, 20, 12$, and 12. Since A_5 is simple, we cannot produce irreducible characters of A_5 by lifting irreducible characters of quotient groups, as we have done in previous examples. However, A_5 is a group of permutations, so we have as a starting point the characters of certain obvious permutation modules.

Let $X = \{1, 2, 3, 4, 5\}$; A_5 acts transitively on X in the obvious way. Let π be the character of the $\mathbb{C}A_5$-module $\mathbb{C}X$. By Example 4, we see that $\pi(g)$ equals the number of fixed points of X under the action of $g \in A_5$, and hence we have

	1	15	20	12	12
	1	(1 2)(3 4)	(1 2 3)	(1 2 3 4 5)	(1 3 5 2 4)
π	5	1	2	0	0
$\pi - \chi_1$	4	0	1	-1	-1

We also have

$$(\pi - \chi_1, \pi - \chi_1) = \frac{1}{|A_5|} \sum_{i=1}^{5} k_i[(\pi - \chi_1)(g_i)]^2$$

$$= \frac{1}{60}(1 \cdot 16 + 15 \cdot 0 + 20 \cdot 1 + 12 \cdot 1 + 12 \cdot 1) = 1,$$

and hence $\pi - \chi_1$ is irreducible by Corollary 3. Let $\chi_2 = \pi - \chi_1$.

Let Y be the set of two-element subsets of X; this is a transitive A_5-set under the obvious action. Let ψ be the character of the $\mathbb{C}A_5$-module $\mathbb{C}Y$. We have

	1	15	20	12	12
	1	(1 2)(3 4)	(1 2 3)	(1 2 3 4 5)	(1 3 5 2 4)
ψ	10	2	1	0	0
$\psi - \chi_1$	9	1	0	-1	-1

Now

$$(\psi - \chi_1, \psi - \chi_1) = \frac{1}{|A_5|} \sum_{i=1}^{5} k_i[(\psi - \chi_1)(g_i)]^2$$

$$= \frac{1}{60}(1 \cdot 81 + 15 \cdot 1 + 20 \cdot 0 + 12 \cdot 1 + 12 \cdot 1) = 2,$$

so by Corollary 3, $\psi - \chi_1$ is the sum of two irreducible characters; as noted in Example 4, neither of these two characters is χ_1, since Y is transitive. We have

$$(\psi - \chi_1, \chi_2) = \frac{1}{|A_5|} \sum_{i=1}^{5} k_i(\psi - \chi_1)(g)\chi_2(g)$$

$$= \frac{1}{60}(1 \cdot 9 \cdot 4 + 15 \cdot 1 \cdot 0 + 20 \cdot 0 \cdot 1 + 24 \cdot (-1)(-1))$$

$$= 1,$$

and therefore by Corollary 4 we see that $\psi - \chi_1 - \chi_2$ is an irreducible character that is neither χ_1 nor χ_2. Let $\chi_3 = \psi - \chi_1 - \chi_2$, and observe that $f_3 = 5$.

By Corollary 14.2 we have $\sum_{i=1}^{5} f_i^2 = |A_5| = 60$, which gives $f_4^2 + f_5^2 = 18$ and hence $f_4 = f_5 = 3$. Thus we have the following partial character table:

	1	15	20	12	12
	1	(1 2)(3 4)	(1 2 3)	(1 2 3 4 5)	(1 3 5 2 4)
χ_1	1	1	1	1	1
χ_2	4	0	1	-1	-1
χ_3	5	1	-1	0	0
χ_4	3				
χ_5	3				

Let $a = \chi_4((1\ 2)(3\ 4))$ and $b = \chi_5((1\ 2)(3\ 4))$. By column orthogonality, we have

$$0 = \sum_{i=1}^{5} f_i \chi_i((1\ 2)(3\ 4)) = 1 \cdot 1 + 4 \cdot 0 + 5 \cdot 1 + 3a + 3b,$$

and hence $a + b = -2$. By part (iii) of Proposition 14.4, we see that each of a and b is a sum of three square roots of unity; therefore $a, b \in \{-3, -1, 1, 3\}$. But by column orthogonality we have

$$4 = \frac{60}{15} = \frac{|A_5|}{k_2} = \sum_{i=1}^{5} |\chi_i((1\ 2)(3\ 4))|^2 = 1 + 0 + 1 + a^2 + b^2,$$

and hence $a^2 + b^2 = 2$. It now follows that $a = b = -1$.

By column orthogonality, we have

$$3 = \frac{60}{20} = \frac{|A_5|}{k_3} = \sum_{i=1}^{5} |\chi_i((1\ 2\ 3))|^2$$
$$= 1 + 1 + 1 + |\chi_4((1\ 2\ 3))|^2 + |\chi_5((1\ 2\ 3))|^2;$$

thus $|\chi_4((1\ 2\ 3))|^2 + |\chi_5((1\ 2\ 3))|^2 = 0$, and therefore we conclude that $\chi_4((1\ 2\ 3)) = \chi_5((1\ 2\ 3)) = 0$. We now have the following partial character table:

	1	15	20	12	12
	1	$(1\ 2)(3\ 4)$	$(1\ 2\ 3)$	$(1\ 2\ 3\ 4\ 5)$	$(1\ 3\ 5\ 2\ 4)$
χ_1	1	1	1	1	1
χ_2	4	0	1	-1	-1
χ_3	5	1	-1	0	0
χ_4	3	-1	0		
χ_5	3	-1	0		

Let $x = (1\ 2\ 3\ 4\ 5)$. Recall that in proving that x and x^2 are not conjugate in A_5, we found that x and x^{-1} are conjugate in A_5; hence $\chi_4(x) = \chi_4(x^{-1})$, and thus $\chi_4(x)$ is real-valued by part (iv) of Proposition 14.4. The same argument shows that $\chi_4(x^2)$, $\chi_5(x)$, and $\chi_5(x^2)$ are also real-valued. Let $c = \chi_4(x)$ and $d = \chi_4(x^2)$. Then by row orthogonality, we have

$$0 = \sum_{i=1}^{5} k_i \chi_4(g_i) = 1 \cdot 3 + 15 \cdot (-1) + 20 \cdot 0 + 12c + 12d,$$

which gives $c + d = 1$, and also (since χ_4 is real-valued)

$$60 = \sum_{i=1}^{5} k_i \chi_4(g_i)^2 = 1 \cdot 9 + 15 \cdot 1 + 20 \cdot 0 + 12c^2 + 12d^2,$$

which gives $c^2 + d^2 = 3$. Hence $c^2 - c - 1 = 0$, so without loss of generality we have $c = \frac{1+\sqrt{5}}{2}$ and $d = 1 - c = \frac{1-\sqrt{5}}{2}$. We can similarly show that we must have $\chi_5(x), \chi_5(x^2) \in \{\frac{1+\sqrt{5}}{2}, \frac{1-\sqrt{5}}{2}\}$; we leave it to the reader to verify that $\chi_5(x) = \frac{1-\sqrt{5}}{2}$ and $\chi_5(x^2) = \frac{1+\sqrt{5}}{2}$, which completes the character table of A_5.

EXERCISES

Throughout these exercises, G denotes a finite group, and $\mathcal{X}$ denotes the character table of G.

1. What is the determinant of $\mathcal{X}$?

2. Show that the sum of the entries in any row of $\mathcal{X}$ is a non-negative integer.

3. Compute the character of the G-set G, where G acts via conjugation, and determine the multiplicities in this character of the irreducible characters.

4. Observe that the complex conjugate of an irreducible character is an irreducible character. (This is either Exercise 14.6 or an easy consequence of part (ii) of Proposition 14.8 and Corollary 3.) Let P be the $r \times r$ permutation matrix such that $P\mathcal{X}$ is the matrix obtained from $\mathcal{X}$ by interchanging two rows if their corresponding characters are conjugate. Let Q be the $r \times r$ permutation matrix such that $\mathcal{X}Q$ is the matrix obtained from $\mathcal{X}$ by interchanging two columns if their corresponding conjugacy classes are inverse, in the sense that one class consists of the inverses of the elements of the other class. Show that $P\mathcal{X} = \mathcal{X}Q$.

5. Suppose that $|G|$ is odd. Show that the only irreducible character that is real-valued is the principal character.

6. Let χ be an irreducible character of G, and let $e \in \mathbb{C}G$ be the identity element of the corresponding matrix summand of $\mathbb{C}G$. Show that

$$e = \frac{\chi(1)}{|G|} \sum_{g \in G} \chi(g^{-1})g.$$

7. If $N \trianglelefteq G$ and $g \in G$, show that $|C_G(g)| \geq |C_{G/N}(gN)|$.

8. Show that if G acts doubly transitively on X, then the character of $\mathbb{C}X$ is the sum of the principal character and exactly one other irreducible character.
9. Determine the character table of D_{2n} for any $n \in \mathbb{N}$.
10. Determine the character table of Σ_5.

16. Induction

Induced modules play a critical role in the representation theory of finite groups. However, even the definition of an induced module offers a pedagogical challenge, one which we address by offering two equivalent definitions. (A third is developed in Exercise 1.) We first present induced modules using tensor products over arbitrary rings; this approach is slick, but it has more algebraic sophistication than is strictly necessary for our purposes. We then go over the same material using an approach that, while clumsy in comparison, involves only a knowledge of tensor products over fields.

Let $H \leqslant G$, and let V be an FH-module. Since the group algebra FG is an (FG, FH)-bimodule, we can construct the FG-module $FG \otimes_{FH} V$. We call this FG-module the *induced FG-module of V* (or the *induction* of V to G), and we denote it by $\operatorname{Ind}_H^G V$. (Other common notations for $FG \otimes_{FH} V$ include V^G, $V \uparrow G$, and $V \uparrow_H^G$.)

LEMMA 1. We have $\dim_F \operatorname{Ind}_H^G V = |G : H| \dim_F V$ for V an FH-module. Moreover, as F-vector spaces we have

$$\operatorname{Ind}_H^G V = \bigoplus_{t \in T} t \otimes V,$$

where T is a transversal for H in G and $t \otimes V = \{t \otimes v \mid v \in V\}$.

PROOF. Let T be a transversal for H in G. Then G is the disjoint union of the sets tH for $t \in T$, and from this it follows that $FG \cong |T|FH$ as FH-modules; if we write $g \in G$ as $g = th$ for a unique choice of $t \in T$ and $h \in H$, then the image of g under this isomorphism is h in the summand corresponding to t (and 0 in all other summands). Using Propositions 12.2 and 12.3, we see that $\operatorname{Ind}_H^G V = FG \otimes_{FH} V \cong (|T|FH) \otimes_{FH} V \cong |T|(FH \otimes_{FH} V) \cong |T|V$ as

FH-modules, and so $\dim_F \operatorname{Ind}_H^G V = |T| \dim_F V = |G:H| \dim_F V$. The second statement follows from the fact that $g \otimes v = t \otimes hv \in t \otimes V$ for any $v \in V$ if $g = th$ is as above.   ∎

Observe that if $t \in T$, then $g \in G$ maps $t \otimes V$ to $s \otimes V$, where $s \in T$ is such that $gt = sh$ for some $h \in H$.

Let U be an FG-module, and let $H \leqslant G$. When regarding U as an FH-module, we will write $\operatorname{Res}_H^G U$ in place of U; this FH-module is called the *restriction* of U to H. (Other common notations for the restriction of U to H include U_H, $U \downarrow H$, and $U \downarrow_H^G$.)

FROBENIUS RECIPROCITY THEOREM. Let U be an FG-module, let $H \leqslant G$, and let V be an FH-module. Then as F-vector spaces we have $\operatorname{Hom}_{FH}(V, \operatorname{Res}_H^G U) \cong \operatorname{Hom}_{FG}(\operatorname{Ind}_H^G V, U)$.

PROOF. Let $\varphi \in \operatorname{Hom}_{FH}(V, \operatorname{Res}_H^G U)$, and consider the mapping $f_\varphi \colon FG \times V \to U$ that sends (g, v) to $g\varphi(v)$. If $h \in H$, then we have $f_\varphi(gh, v) = gh\varphi(v) = g\varphi(hv) = f_\varphi(g, hv)$; from this, we see easily that f_φ is balanced. Therefore we obtain an element $\Gamma(\varphi)$ of $\operatorname{Hom}_{FG}(\operatorname{Ind}_H^G V, U)$ sending $g \otimes v$ to $g\varphi(v)$, and hence we get a map $\Gamma \colon \operatorname{Hom}_{FH}(V, \operatorname{Res}_H^G U) \to \operatorname{Hom}_{FG}(\operatorname{Ind}_H^G V, U)$. We easily see that Γ is linear, and also that $\Gamma(\varphi) = 0$ implies $\varphi = 0$ and hence that Γ is injective. If $\theta \colon \operatorname{Ind}_H^G V \to U$ is an FG-module homomorphism, then we define an FH-module homomorphism $\varphi \colon V \to \operatorname{Res}_H^G U$ by $\varphi(v) = \theta(1 \otimes v)$; we have $\Gamma(\varphi)(g \otimes v) = g\varphi(v) = g\theta(1 \otimes v) = \theta(g \otimes v)$, showing that $\Gamma(\varphi) = \theta$ and hence that Γ is surjective.   ∎

Again let V be an FH-module; we now give an alternate description of the induction of V to G. Considering FG and V as F-vector spaces, we can construct the F-vector space $FG \otimes_F V$. We give $FG \otimes_F V$ an FG-module structure by considering V as an FG-module on which G acts trivially, so that for $g, x \in G$ and $v \in V$ we define $x(g \otimes v) = xg \otimes v$. Let Y be the subspace of $FG \otimes_F V$ that is spanned by $\{gh \otimes v - g \otimes hv \mid g \in G,\ h \in H,\ v \in V\}$, and let $\operatorname{Ind}_H^G V$ be the quotient space $FG \otimes_F V / Y$. If $g \in G$ and $v \in V$, then we still use $g \otimes v$ for the image in $\operatorname{Ind}_H^G V$ of $g \otimes v \in FG \otimes_F V$, so that in $\operatorname{Ind}_H^G V$ we have $gh \otimes v = g \otimes hv$ for any $h \in H$. We have $x(gh \otimes v - g \otimes hv) = (xg)h \otimes v - (xg) \otimes hv$ for any $g, x \in G$, $h \in H$, and $v \in V$; consequently, Y is an FG-submodule of $FG \otimes_F V$. Hence $\operatorname{Ind}_H^G V$ is an FG-module, with the action of $x \in G$ on $g \otimes v \in \operatorname{Ind}_H^G V$ given by $x(g \otimes v) = xg \otimes v$.

Now let T be a transversal for H in G, and let $\{v_i\}_{i \in I}$ be an F-basis for V. It is not hard to show that the subspace Y is spanned by $\{th \otimes v_i - t \otimes hv_i \mid t \in T,\, i \in I\}$. This set has $|T|(|H| - 1)\dim_F V$ elements, and so $\dim_F \operatorname{Ind}_H^G V \geq |G : H|\dim_F V$. But as any $g \otimes v$ can be written as an F-linear combination of the $t \otimes v_i$, this inequality must be an equality, which establishes Lemma 1 for this definition of induced modules. We leave it to the reader to construct a proof of Frobenius reciprocity for this definition of induction. (Our two definitions are in fact equivalent, for the FG-modules $FG \otimes_F V/Y$ and $FG \otimes_{FH} V$ are easily seen to be isomorphic.)

We now restrict our attention to the case $F = \mathbb{C}$. If V is a $\mathbb{C}H$-module having character φ, then we denote the character of the $\mathbb{C}G$-module $\operatorname{Ind}_H^G V$ by φ^G, and we call φ^G an *induced character*. We have $\varphi^G(1) = |G : H|\varphi(1)$ by Lemma 1. Similarly, if U is a $\mathbb{C}G$-module having character χ, then we denote the character of the $\mathbb{C}H$-module $\operatorname{Res}_H^G U$ by $\chi|_H$. Observe that $\chi|_H(1) = \chi(1)$.

The following result is the form of "Frobenius reciprocity" that was actually known to Frobenius.

THEOREM 2. Let $H \leqslant G$, let U be a $\mathbb{C}G$-module having character χ, and let V be a $\mathbb{C}H$-module having character φ. Then $(\varphi, \chi|_H)_H = (\varphi^G, \chi)_G$.

(Here, for any group K, we use $(\ ,\ )_K$ to denote the inner product on the space of complex-valued class functions on K.)

PROOF. This follows immediately from the Frobenius reciprocity theorem and Theorem 14.12. ∎

If $H \leqslant G$, then Theorem 2 enables us to express characters induced from H to G in terms of the irreducible characters of G, as follows. Let $\varphi_1, \ldots, \varphi_r$ and $\chi_1, \ldots, \chi_s$ be the irreducible characters of H and G, respectively. We define the *induction-restriction table* of (G, H) to be the $r \times s$ matrix whose (i, j)-entry gives the common value of $(\varphi_i^G, \chi_j)_G$ and $(\varphi_i, \chi_j|_H)_H$. Using Corollary 15.4, we see that the ith row of this table gives the multiplicities with which the χ_j appear in the decomposition of φ_i^G, while the jth column gives the multiplicities with which the φ_i appear in the decomposition of $\chi_j|_H$. We can compute the table by calculating the restrictions of the χ_j and then read off the expressions for the φ_i^G in terms of the χ_j.

For example, let $G = \Sigma_3$ and $H = \Sigma_2 \leqslant G$. The character table of H is

	1	1
	1	(1 2)
φ_1	1	1
φ_2	1	-1

while that of G is

	1	3	2
	1	(1 2)	(1 2 3)
χ_1	1	1	1
χ_2	1	-1	1
χ_3	2	0	-1

By inspection, we have $\chi_1|_H = \varphi_1$, $\chi_2|_H = \varphi_2$, and $\chi_3|_H = \varphi_1 + \varphi_2$. Hence the induction-restriction table is

	χ_1	χ_2	χ_3
φ_1	1	0	1
φ_2	0	1	1

Therefore, $\varphi_1^G = \chi_1 + \chi_3$ and $\varphi_2^G = \chi_2 + \chi_3$.

Constructing the induction-restriction table of (G, H) requires that we know in advance the character table of G. A more common circumstance is that we know the character table of H but not that of G and we wish to use induction from H to G as a means of producing characters of G. Our next task is to derive formulas that allow us to explicitly calculate induced characters.

PROPOSITION 3. Let $H \leqslant G$, and let V be a $\mathbb{C}H$-module having character χ. If T is a transversal for H in G, then for any $g \in G$ we have

$$\chi^G(g) = \sum_{\substack{t \in T, \\ t^{-1}gt \in H}} \chi(t^{-1}gt).$$

PROOF. Let $g \in G$. From Lemma 1, we have $\mathrm{Ind}_H^G V = \oplus_{t \in T} t \otimes V$. Fix a $\mathbb{C}$-basis $\{v_i\}$ of V, so that $\{t \otimes v_i \mid t \in T\}$ is a $\mathbb{C}$-basis of $\mathrm{Ind}_H^G V$. Recall that $\chi^G(g)$ is the trace of the transformation of $\mathrm{Ind}_H^G V$ defined by g. Let $t \in T$; then if $s \in T$ and $h \in H$ are such that $gt = sh$, then we have $g(t \otimes V) = s \otimes hV \subseteq s \otimes V$. Consequently, if $s \neq t$, or equivalently if $t^{-1}gt \notin H$, then the $t \otimes V$ component makes no contribution to $\chi^G(g)$. However, if $s = t$, or equivalently if $t^{-1}gt \in H$, then we have $g(t \otimes V) = tt^{-1}gt \otimes V = t \otimes t^{-1}gtV$, which shows that the linear transformation of $t \otimes V$ defined by g is the same as that defined by $t^{-1}gt$ on V. Thus in this case, the contribution made to $\chi^G(g)$ by the $t \otimes V$ component is $\chi(t^{-1}gt)$. The result now follows. ∎

COROLLARY 4. Let χ be a character of $H \leqslant G$, and let $g \in G$. Then

$$\chi^G(g) = \frac{1}{|H|} \sum_{\substack{x \in G, \\ x^{-1}gx \in H}} \chi(x^{-1}gx).$$

PROOF. Let T be a transversal for H in G. Let $x \in G$, and write $x = th$, where $t \in T$ and $h \in H$. Then $x^{-1}gx = h^{-1}(t^{-1}gt)h$; thus $x^{-1}gx \in H$ iff $t^{-1}gt \in H$, and we have $\chi(x^{-1}gx) = \chi(t^{-1}gt)$ in this case since χ is a class function. It follows that for every $t \in T$ such that $t^{-1}gt \in H$, there exist $|H|$ elements $x \in G$ such that $x^{-1}gx \in H$, and for these x we have $\chi(x^{-1}gx) = \chi(t^{-1}gt)$; the result now follows from Proposition 3. ∎

PROPOSITION 5. Let χ be a character of $H \leqslant G$, and let $g \in G$, and let s be the number of conjugacy classes of H whose members are conjugate in G to g. If $s = 0$, then $\chi^G(g) = 0$. Otherwise, if we let $h_1, \ldots, h_s$ be representatives of these s conjugacy classes of H, then

$$\chi^G(g) = \sum_{i=1}^{s} \frac{|C_G(g)|}{|C_H(h_i)|} \chi(h_i).$$

PROOF. If $s = 0$, then $\{x \in G \mid x^{-1}gx \in H\}$ is the empty set, and it follows from Corollary 4 that $\chi^G(g) = 0$. Assume that $s > 0$, and let $X_i = \{x \in G \mid x^{-1}gx$ lies in H and is conjugate in H to $h_i\}$ for each $1 \leq i \leq s$. The sets X_i are pairwise disjoint, and their union is $\{x \in G \mid x^{-1}gx \in H\}$. Therefore by Corollary 4 we have

$$\chi^G(g) = \frac{1}{|H|} \sum_{\substack{x \in G, \\ x^{-1}gx \in H}} \chi(x^{-1}gx)$$

$$= \frac{1}{|H|} \sum_{i=1}^{s} \sum_{x \in X_i} \chi(x^{-1}gx)$$

$$= \frac{1}{|H|} \sum_{i=1}^{s} \sum_{x \in X_i} \chi(h_i) = \sum_{i=1}^{s} \frac{|X_i|}{|H|} \chi(h_i).$$

Fix some $1 \leq i \leq s$, and choose some $t_i \in G$ such that $t_i^{-1}gt_i = h_i$. Then for any $c \in C_G(g)$ and $h \in H$, we have

$$(ct_ih)^{-1}g(ct_ih) = h^{-1}t_i^{-1}c^{-1}gct_ih = h^{-1}t_i^{-1}c^{-1}cgt_ih$$
$$= h^{-1}t_i^{-1}gt_ih = h^{-1}h_ih,$$

which shows that $ct_ih \in X_i$ and hence that $C_G(g)t_iH \subseteq X_i$. Conversely, if $x \in X_i$, then we have $x^{-1}gx = h^{-1}h_ih = h^{-1}(t_i^{-1}gt_i)h$ for some $h \in H$; thus $xh^{-1}t_i^{-1} \in C_G(g)$, and therefore we have $x \in C_G(g)t_ih \subseteq C_G(g)t_iH$, giving $X_i = C_G(g)t_iH$. By Proposition 3.15, we now have

$$|X_i| = |C_G(g)t_iH| = \frac{|C_G(g)||H|}{|H \cap t_i^{-1}C_G(g)t_i|}.$$

But $t_i^{-1}C_G(g)t_i = C_G(t_i^{-1}gt_i) = C_G(h_i)$ by Exercise 3.7, and therefore $|X_i| = |H : H \cap C_G(h_i)||C_G(g)| = |H : C_H(h_i)||C_G(g)|$. Thus

$$\frac{|X_i|}{|H|} = \frac{|H : C_H(h_i)||C_G(g)|}{|H|} = \frac{|C_G(g)|}{|C_H(h_i)|}$$

for each $1 \leq i \leq s$, which completes the proof. ∎

COROLLARY 6. Let χ be a character of $H \leq G$. Let $g \in G$, and suppose that the number s of conjugacy classes of H whose members are conjugate in G to g is positive. Let ℓ be the order of the conjugacy class in G of g; let $h_1, \ldots, h_s$ be representatives of these s conjugacy classes of H, and let $k_1, \ldots, k_s$ be the orders of these classes. Then

$$\chi^G(g) = \sum_{i=1}^{s} |G : H|\frac{k_i}{\ell}\chi(h_i).$$

PROOF. This follows directly from Proposition 3.13 and Proposition 5. ∎

For example, let $G = \Sigma_4$ and $H = \Sigma_3 \leq G$. The character table of H is

	1	3	2
	1	(1 2)	(1 2 3)
φ_1	1	1	1
φ_2	1	-1	1
φ_3	2	0	-1

We wish to find the induced characters $\varphi_1^G, \varphi_2^G, \varphi_3^G$ via the formula given in Corollary 6. Recall from Example 15.7 that Σ_4 has 5 conjugacy classes, with representatives $1, (1\ 2)(3\ 4), (1\ 2\ 3), (1\ 2), (1\ 2\ 3\ 4)$ and respective orders $1, 3, 8, 6,$ and 6. Using Proposition 1.10, we see that $(1\ 2)(3\ 4)$ and $(1\ 2\ 3\ 4)$ have no conjugates that lie in H, that the identity element has exactly one conjugate lying in H, that $(1\ 2)$ has 3 conjugates in H, and that $(1\ 2\ 3)$ has 2 conjugates in H. Therefore, for each $1 \le i \le 3$, we have

$$\varphi_i^G(1) = 4 \cdot \frac{1}{1} \cdot \varphi_i(1) = 4\varphi_i(1),$$

$$\varphi_i^G((1\ 2)(3\ 4)) = 0,$$

$$\varphi_i^G((1\ 2\ 3)) = 4 \cdot \frac{2}{8} \cdot \varphi_i((1\ 2\ 3)) = \varphi_i((1\ 2\ 3)),$$

$$\varphi_i^G((1\ 2)) = 4 \cdot \frac{3}{6} \cdot \varphi_i((1\ 2)) = 2\varphi_i((1\ 2)),$$

$$\varphi_i^G((1\ 2\ 3\ 4)) = 0.$$

We conclude that

	1	3	8	6	6
	1	$(1\ 2)(3\ 4)$	$(1\ 2\ 3)$	$(1\ 2)$	$(1\ 2\ 3\ 4)$
φ_1^G	4	0	1	2	0
φ_2^G	4	0	1	−2	0
φ_3^G	8	0	−1	0	0

We end this section by presenting an important theorem on finite groups which, quite remarkably, has no known proof that does not use character theory.

FROBENIUS' THEOREM. Let G be a transitive permutation group on a finite set X, and suppose that each non-identity element of G fixes at most one element of X. Then the union of the identity element and the set of elements of G that have no fixed points is a normal subgroup of G.

(Observe that G is finite, being isomorphic with a subgroup of the finite group Σ_X.)

PROOF. Let $N = \{1\} \cup \{g \in G \mid gx \ne x \text{ for all } x \in X\}$; we wish to show that $N \trianglelefteq G$. Let $|X| = n$, and let $H = G_x$ for some $x \in X$. By Lemma 3.2, the conjugates of H are the stabilizers of single elements of X. By hypothesis, no two of these conjugates can share a non-identity element. It now follows that H has n distinct conjugates and that G has $n(|H| - 1)$ elements that fix exactly one element of X. But $|G| = |X||H| = n|H|$, since X and G/H are isomorphic G-sets by Proposition 3.4, and therefore $|N| = |G| - n(|H| - 1) = n$.

Let $1 \neq h \in H$. Suppose that $h = gh'g^{-1}$ for some $g \in G$ and $h' \in H$. Then h lies in both $H = G_x$ and $gHg^{-1} = G_{gx}$; by hypothesis, this forces $gx = x$, and hence $g \in H$. Therefore, the conjugacy class in G of h is precisely the conjugacy class in H of h. Similarly, if $g \in C_G(h)$, then $h = ghg^{-1} \in G_{gx}$, and hence $g \in H$, which implies that $C_G(h) = C_H(h)$.

Every element of G either lies in N or lies in one of the n stabilizers, each of which is conjugate with H. In other words, every element of G that does not lie in N is conjugate with a non-identity element of H. We conclude that $\{1, h_2, \ldots, h_s, y_1, \ldots, y_t\}$ is a set of conjugacy class representatives for G, where $\{1, h_2, \ldots, h_s\}$ are representatives of the conjugacy classes of H and $\{y_1, \ldots, y_t\}$ are representatives of the conjugacy classes of G which comprise $N - \{1\}$.

Let χ_1 be the principal character of G, and let $\varphi_1, \ldots, \varphi_s$ be the irreducble characters of H. Fix some $1 \leq i \leq s$, and consider the induced character φ_i^G. We have $\varphi_i^G(1) = |G : H|\varphi_i(1) = n\varphi_i(1)$. By Proposition 5, we have $\varphi_i^G(h_j) = \varphi_i(h_j)$ for each $2 \leq j \leq s$ since $C_G(h_j) = C_H(h_j)$, and $\varphi_i^G(y_k) = 0$ for each $1 \leq k \leq t$.

Now fix some $2 \leq i \leq s$, and let $\chi_i = \varphi_i^G - \varphi_i(1)\varphi_1^G + \varphi_i(1)\chi_1$, so that χ_i is a virtual character of G. For any $2 \leq j \leq s$ and $1 \leq k \leq t$, we have

	1	h_j	y_k
φ_i^G	$n\varphi_i(1)$	$\varphi_i(h_j)$	0
$\varphi_i(1)\varphi_1^G$	$n\varphi_i(1)$	$\varphi_i(1)$	0
$\varphi_i(1)\chi_1$	$\varphi_i(1)$	$\varphi_i(1)$	$\varphi_i(1)$
χ_i	$\varphi_i(1)$	$\varphi_i(h_j)$	$\varphi_i(1)$

Therefore,

$$(\chi_i, \chi_i) = \frac{1}{|G|} \sum_{g \in G} |\chi_i(g)|^2 = \frac{1}{|G|} \left(\sum_{g \in N} |\chi_i(g)|^2 + \sum_{x \in X} \sum_{1 \neq g \in G_x} |\chi_i(g)|^2 \right)$$

$$= \frac{1}{|G|} \left(n\varphi_i(1)^2 + n \sum_{1 \neq h \in H} |\chi_i(h)|^2 \right)$$

$$= \frac{1}{|H|} \sum_{h \in H} |\varphi_i(h)|^2 = (\varphi_i, \varphi_i) = 1$$

by row orthogonality. Hence by Corollary 15.2, either χ_i or $-\chi_i$ is an irreducible character of G; since $\chi_i(1) > 0$, we conclude that it is χ_i and not $-\chi_i$ that is an actual character.

Let χ be the character $\sum_{i=1}^{s} \chi_i(1)\chi_i$ of G. Column orthogonality implies that $\chi(h) = \sum_{i=1}^{s} \varphi_i(1)\varphi_i(h) = 0$ for any $1 \neq h \in H$, and for any $y \in N$ we see via Corollary 14.2 that $\chi(y) = \sum_{i=1}^{s} \varphi_i(1)^2 = |H|$. It now follows that $\chi(g) = |H|$ if $g \in N$ and $\chi(g) = 0$ if $g \in G - N$, and hence that $N = \{g \in G \mid \chi(g) = \chi(1)\}$. Therefore $N \trianglelefteq G$ by part (vi) of Proposition 14.4. (Having established this, we now see that this character χ is in fact the lift to G of the regular character of G/N.) ∎

(The method of proof we have used above is an example of the method of *exceptional characters*; see [15, Chapter 7].)

A *Frobenius group* is a group G having a subgroup H such that $H \cap gHg^{-1} = 1$ for every $g \in G - H$. We call H a *Frobenius complement* of G. If G is a Frobenius group with Frobenius complement H, then the action of G on G/H is transitive and faithful. Furthermore, if $1 \neq g \in G$ fixes both xH and yH, then we find that $g \in xHx^{-1} \cap yHy^{-1}$; this implies that $H \cap (y^{-1}x)H(y^{-1}x)^{-1} \neq 1$, which forces $xH = yH$. Therefore, any finite Frobenius group G satisfies the hypothesis of Frobenius' theorem and hence has a normal subgroup N, called the *Frobenius kernel* of G. Since H is the stabilizer of $H \in G/H$ under the action of G, and since the elements of N by definition fix no elements of G/H, we must have $N \cap H = 1$; furthermore, as argued in the first paragraph of the proof of Frobenius' theorem, we have $|N||H| = |G|$. Therefore $G = N \rtimes H$.

Let G be a finite Frobenius group with Frobenius kernel N and Frobenius complement H, and let $\varphi \colon H \to \operatorname{Aut}(N)$ be the conjugation homomorphism of $G = N \rtimes H$, so that $\varphi(h)(n) = hnh^{-1}$ for $h \in H$ and $n \in N$. If $1 \neq h \in H$, then

$$C_G(h) \cap N = \{n \in N \mid nh = hn\} = \{n \in N \mid h = n^{-1}\varphi(h)(n)h\}$$
$$= \{n \in N \mid \varphi(h)(n) = n\},$$

and so $C_G(h) \cap N$ is the set of fixed points of $\varphi(h) \in \operatorname{Aut}(N)$. However, recall that in the proof of Frobenius' theorem we observed that $C_G(h) = C_H(h) \leqslant H$; since $N \cap H = 1$, we now have $C_G(h) \cap N = 1$. Therefore for every $1 \neq h \in H$, $\varphi(h)$ is a *fixed-point-free automorphism* of N, meaning that if $\varphi(h)(n)$ for $n \in N$, then we must have $n = 1$. (In particular, φ is injective.)

Conversely, let N and H be finite groups, and let $\varphi\colon H \to \operatorname{Aut}(N)$ be a homomorphism for which every $\varphi(h)$ is fixed-point-free. Let $n \in N$, and let $h \in H \cap nHn^{-1}$. Then $h = nh'n^{-1}$ for some $h' \in H$, so $nh'h^{-1} = hnh^{-1} = \varphi(h)(n) \in N$ and hence $h'h^{-1} \in N \cap H = 1$. Therefore $h = h'$, so we now have $n = \varphi(h)(n)$, which since $\varphi(h)$ is fixed-point-free forces either $n = 1$ or $h = 1$. Now if $g \in G - H$, then $g = nh$ for some $1 \neq n \in N$ and some $h \in H$, and so by the above we see that $H \cap gHg^{-1} = H \cap nHn^{-1} = 1$. Therefore $N \rtimes_\varphi H$ is a Frobenius group. For example, if p and q are distinct primes with $p \equiv 1 \pmod{q}$, then we find that any monomorphism $\varphi\colon \mathbf{Z_q} \to \operatorname{Aut}(\mathbf{Z_p})$ is fixed-point-free, and hence the unique non-abelian group of order pq is a Frobenius group.

A theorem of Thompson states that any finite group having a fixed-point-free automorphism of prime order is nilpotent. This implies that the Frobenius kernel of a finite Frobenius group is nilpotent. (See the discussion in [10, Section 1.6].) This was first established by Thompson in his 1959 Ph.D. thesis at the University of Chicago. Both the result itself and the nature of his proof, which was perhaps the first appearance in group theory of an intricate argument spanning dozens of pages, were of fundamental importance in the development of finite group theory in the subsequent decades.

EXERCISES

Throughout these exercises, G is a finite group, H is a subgroup of G, F is a field, and all FG-modules are finitely generated.

1. Let V be an FH-module, and let W be an FG-module containing V. Suppose that W has the property that for any FG-module U and any $\varphi \in \operatorname{Hom}_{FH}(V, U)$, there is a unique FG-module homomorphism from W to U which extends U. (In this case, we say that W is *relatively H-free* with respect to V; compare with Exercises 13.1–2.) Show that $W \cong \operatorname{Ind}_H^G V$.

2. Let W be an FG-module that is generated by an FH-submodule V, and suppose that $\dim_F W = |G : H| \dim_F V$. Show that we must have $W \cong \operatorname{Ind}_H^G V$.

3. Show that $\operatorname{Hom}_{FG}(U, \operatorname{Ind}_H^G V) \cong \operatorname{Hom}_{FH}(\operatorname{Res}_H^G U, V)$ as F-vector spaces for any FG-module U and FH-module V. (HINT: Show first that if $\varphi \in \operatorname{Hom}_{FH}(\operatorname{Res}_H^G U, V)$, then the map sending $u \in U$ to $\sum_{t \in T} t \otimes \varphi(t^{-1}u)$, where T is a transversal for H in G, lies in $\operatorname{Hom}_{FG}(U, \operatorname{Ind}_H^G V)$.)

4. Show that if X is a transitive G-set, then $FX \cong \mathrm{Ind}_{G_x}^{G} F$ for any $x \in X$.

5. Compute the induction-restriction table of (A_5, A_4).

6. Compute the character table of the unique non-abelian group of order pq, where p and q are distinct primes and $p \equiv 1 \pmod{q}$.

7. Suppose that G is a Frobenius group having Frobenius kernel N. Show that the irreducible characters of G either are lifts of irreducible characters of G/N or are induced from the non-principal irreducible characters of N. When is it true that the inductions to G of two distinct non-principal irreducible characters of N are equal?

FURTHER EXERCISES

In this set of exercises, we develop the character table of $G = \mathrm{GL}(2, q)$, where q is any prime power. Recall from Proposition 4.2 that we have $|G| = q(q-1)^2(q+1)$.

Our first task is to determine the conjugacy classes of G. Here we sketch a comparatively "hands-on" approach, which uses only some standard topics from linear algebra. (A slicker approach might, for instance, make use of the invariant factor formulation of rational canonical form.) We denote by $m_g(X)$ the minimal polynomial of an element $g \in G$. Our strategy is to exploit the following:

8. Show that conjugate elements of G have the same minimal polynomial.

Since $m_g(X)$ is a factor of the characteristic polynomial of g, which is the quadratic polynomial $\det(XI - g)$, we see that $m_g(X)$ is either linear, or a product of two like or unlike linear factors, or an irreducible monic quadratic.

9. (cont.) Show that if $m_g(X) = X - a$ for some $a \in \mathbb{F}_q^{\times}$, then $g = \left(\begin{smallmatrix} a & 0 \\ 0 & a \end{smallmatrix}\right) \in Z(G)$.

10. (cont.) Show that if $m_g(X) = (X - a)^2$ for some $a \in \mathbb{F}_q^{\times}$, then g is conjugate with $\left(\begin{smallmatrix} a & 1 \\ 0 & a \end{smallmatrix}\right)$, and the conjugacy class of g has order $q^2 - 1$.

11. (cont.) Show that if $m_g(X) = (X - a)(X - b)$ for some distinct $a, b \in \mathbb{F}_q^{\times}$, then g is conjugate with $\left(\begin{smallmatrix} a & 0 \\ 0 & b \end{smallmatrix}\right)$, and the conjugacy class of g has order $q^2 + q$.

12. (cont.) Show that if $r, s \in \mathbb{F}_q$ are such that $X^2 - rX - s$ is irreducible, then the conjugacy class of $\left(\begin{smallmatrix} 0 & 1 \\ s & r \end{smallmatrix}\right)$ has order $q^2 - q$.

(Hint for Exercises 10–12: Compute centralizers and use Proposition 3.13.)

13. (cont.) Verify that what follows is a complete list of the conjugacy classes of G: $q - 1$ classes of 1 element each, indexed by elements

of $\mathbb{F}_q^\times$; $q - 1$ classes of $q^2 - 1$ elements each, indexed by elements of $\mathbb{F}_q^\times$; $(q-1)(q-2)/2$ classes of $q^2 + q$ elements each, indexed by unordered pairs of distinct elements of $\mathbb{F}_q^\times$; and $q(q-1)/2$ classes of $q^2 - q$ elements each, indexed by irreducible monic quadratic polynomials with coefficients in $\mathbb{F}_q$.

Having determined the classes of G, we now turn to finding irreducible characters. Recall that we have an epimorphism $\det: G \to \mathbb{F}_q^\times$. Consequently, we can lift characters of $\mathbb{F}_q^\times$ to characters of G, as in Lemma 15.6. Now $\mathbb{F}_q^\times$ is an abelian group of order $q - 1$, and hence it has $q - 1$ linear characters, which we denote by $\tau_1, \dots, \tau_{q-1}$. (In fact, it is true (see [17, p. 132]) that $\mathbb{F}_q^\times \cong \mathbf{Z}_{q-1}$, and so we can, upon fixing a generator of $\mathbb{F}_q^\times$, completely specify the τ_i.) For each $1 \leq i \leq q - 1$, let $\chi_i = \tau_i \circ \det$. We have

	$\left(\begin{smallmatrix} a & 0 \\ 0 & a \end{smallmatrix}\right)$	$\left(\begin{smallmatrix} a & 1 \\ 0 & a \end{smallmatrix}\right)$	$\left(\begin{smallmatrix} a & 0 \\ 0 & b \end{smallmatrix}\right)$	$\left(\begin{smallmatrix} 0 & 1 \\ s & r \end{smallmatrix}\right)$
χ_i	$\tau_i(a)^2$	$\tau_i(a)^2$	$\tau_i(a)\tau_i(b)$	$\tau_i(-s)$

The characters χ_i are linear, and χ_1 is indeed the principal character of G.

As in Chapter 2, let B, U, and T be the subgroups of G consisting, respectively, of all upper triangular, all upper unitriangular, and all diagonal elements of G. We have $T \cong \mathbb{F}_q^\times \times \mathbb{F}_q^\times$, and so we see from Example 15.2 that the irreducible characters of T have the form $\tau_i \times \tau_j$ (after making appropriate identifications). Now $B = U \rtimes T$ by Proposition 5.1, so we can lift each character $\tau_i \times \tau_j$ of $T \cong B/U$ to obtain a character θ_{ij} of B. The epimorphism from B to T sends $\left(\begin{smallmatrix} x & y \\ 0 & z \end{smallmatrix}\right)$ to $\left(\begin{smallmatrix} x & 0 \\ 0 & z \end{smallmatrix}\right)$, and consequently we have $\theta_{ij}\left(\begin{smallmatrix} x & y \\ 0 & z \end{smallmatrix}\right) = (\tau_i \times \tau_j)\left(\begin{smallmatrix} x & 0 \\ 0 & z \end{smallmatrix}\right) = \tau_i(x)\tau_j(z)$. We now wish to consider the induced characters θ_{ij}^G.

14. (cont.) Show that

	$\left(\begin{smallmatrix} a & 0 \\ 0 & a \end{smallmatrix}\right)$	$\left(\begin{smallmatrix} a & 1 \\ 0 & a \end{smallmatrix}\right)$	$\left(\begin{smallmatrix} a & 0 \\ 0 & b \end{smallmatrix}\right)$	$\left(\begin{smallmatrix} 0 & 1 \\ s & r \end{smallmatrix}\right)$
θ_{ij}^G	$(q+1)\tau_i(a)\tau_j(a)$	$\tau_i(a)\tau_j(a)$	$\tau_i(a)\tau_j(b) + \tau_i(b)\tau_j(a)$	0

Observe that $\theta_{ij}^G = \theta_{ji}^G$.

15. (cont.) Show that $(\theta_{ij}^G, \theta_{ij}^G) = 1 + \delta_{ij}$ and that $(\theta_{ii}^G, \chi_i) = 1$. Conclude that $\theta_{ii}^G - \chi_i$ is an irreducible character of G for each i and that θ_{ij}^G is an irreducible character of G whenever i and j are distinct.

At this point, we have constructed $q - 1$ distinct irreducible characters of degree 1 (the χ_i), $q - 1$ distinct irreducible characters of degree q (the $\theta_{ii}^G - \chi_i$), and $(q-1)(q-2)/2$ distinct irreducible characters of degree $q+1$

(the θ_{ij}^G for $i < j$). Our partial character table of G is now

	$\begin{pmatrix} a & 0 \\ 0 & a \end{pmatrix}$	$\begin{pmatrix} a & 1 \\ 0 & a \end{pmatrix}$	$\begin{pmatrix} a & 0 \\ 0 & b \end{pmatrix}$	$\begin{pmatrix} 0 & 1 \\ s & r \end{pmatrix}$
χ_i	$\tau_i(a)^2$	$\tau_i(a)^2$	$\tau_i(a)\tau_i(b)$	$\tau_i(-s)$
$\theta_{ii}^G - \chi_i$	$q\tau_i(a)^2$	0	$\tau_i(a)\tau_i(b)$	$-\tau_i(-s)$
θ_{ij}^G	$(q+1)\tau_i(a)\tau_j(a)$	$\tau_i(a)\tau_j(a)$	$\tau_i(a)\tau_j(b) + \tau_i(b)\tau_j(a)$	0

To complete the character table, it will be convenient to work with a different set of representatives for the conjugacy classes of those elements whose minimal polynomials are irreducible quadratics. (We shall refer to these classes as being the "fourth-column" classes, owing to their placement in our partial character tables above.) By Exercise 4.4, G has a subgroup C which is isomorphic with $\mathbb{F}_{q^2}^{\times}$, where $\mathbb{F}_{q^2}$ is the field of q^2 elements. As we remarked before, we have $C \cong \mathbf{Z}_{q^2-1}$. It is true, although we are not in a position to prove it, that any subgroup of G that is isomorphic with $\mathbb{F}_{q^2}^{\times}$ is a conjugate of C. (The proof of this fact requires some familiarity with field theory, but the essential point is that all fields of order q^2 are isomorphic; see [19, Corollary XIV.2.7].) It is also true that C must contain $Z(G) \cong \mathbb{F}_q^{\times}$, and so we will consider $\mathbb{F}_q^{\times}$ as being a subgroup of C.

16. (cont.) Show that the elements of $C - \mathbb{F}_q^{\times}$ form a double set of representatives of the fourth-column classes, with $x \in C - \mathbb{F}_q^{\times}$ being conjugate with (but unequal to) x^q. (HINT: Argue that each element of G whose minimal polynomial is an irreducible quadratic determines a subgroup of G that is isomorphic with $\mathbb{F}_{q^2}^{\times}$.)

We now take $x \in C - \mathbb{F}_q^{\times}$ as our arbitrary fourth-column class representative. We have $\chi_i(x) = \tau_i(\det x)$, $(\theta_{ii}^G - \chi_i)(x) = -\tau_i(\det x)$, and $\theta_{ij}^G(x) = 0$.

Let $\beta_1, \ldots, \beta_{q^2-1}$ be the irreducible characters of C. (By choosing a generator for C, we can specify the β_i.) We will consider the induced characters β_k^G.

17. (cont.) Show that

	$\begin{pmatrix} a & 0 \\ 0 & a \end{pmatrix}$	$\begin{pmatrix} a & 1 \\ 0 & a \end{pmatrix}$	$\begin{pmatrix} a & 0 \\ 0 & b \end{pmatrix}$	$x \in C - \mathbb{F}_q^{\times}$
β_k^G	$(q^2 - q)\beta_k(a)$	0	0	$\beta_k(x) + \beta_k(x)^q$

Observe that $(\beta_k^q)^G = \beta_k^G$, where by β_k^q we mean the linear character of C defined by $\beta_k^q(x) = \beta_k(x)^q$.

It is not hard to show that there are exactly $q - 1$ values of k for which $\beta_k^q = \beta_k$. (This follows, for instance, from the observation made in Example 15.1 that the linear characters of a cyclic group of order n themselves form a cyclic group of order n.)

Choose some k such that $\beta_k^q \neq \beta_k$, and let i_k be such that the restriction of β_k to $\mathbb{F}_q^\times \leqslant C$ is τ_{i_k}. Let $\psi_k = (\theta_{11}^G - \chi_1)\theta_{i_k 1}^G - \theta_{i_k 1}^G - \beta_k^G$; this is a virtual character.

18. (cont.) Show that

	$\begin{pmatrix} a & 0 \\ 0 & a \end{pmatrix}$	$\begin{pmatrix} a & 1 \\ 0 & a \end{pmatrix}$	$\begin{pmatrix} a & 0 \\ 0 & b \end{pmatrix}$	$x \in C - \mathbb{F}_q^\times$
ψ_k	$(q-1)\tau_{i_k}(a)$	$\tau_{i_k}(a)$	0	$-\beta_k(x) - \beta_k(x)^q$

19. (cont.) If k is such that $\beta_k^q \neq \beta_k$, show that $(\psi_k, \psi_k) = 1$. Conclude that we have constructed $q(q-1)/2$ distinct irreducible characters of degree $q - 1$.

The number of irreducible characters that we have constructed is now equal to the number of conjugacy classes of G, and so by Theorem 14.3 we are done.

We commented in Section 14 that, in general, there is no natural way of associating a conjugacy class of a group to each irreducible character of that group, even though there are exactly as many conjugacy classes as irreducible characters. However, we do have a very natural correspondence between the irreducible characters and conjugacy classes of GL$(2, q)$, as follows. Let w be a generator of $\mathbb{F}_q^\times$, and let x be a generator of C. Let τ_2 and β_2 be generators of the groups of characters of $\mathbb{F}_q^\times$ and C, respectively, and number the characters so that $\tau_i = \tau_2^{i-1}$ for each i and $\beta_k = \beta_2^{k-1}$ for each k. Then we have the following correspondence between irreducible characters and conjugacy classes:

character	class
χ_i	$\begin{pmatrix} w^{i-1} & 0 \\ 0 & w^{i-1} \end{pmatrix}$
$\theta_{ii}^G - \chi_i$	$\begin{pmatrix} w^{i-1} & 1 \\ 0 & w^{i-1} \end{pmatrix}$
θ_{ij}^G	$\begin{pmatrix} w^{i-1} & 0 \\ 0 & w^{j-1} \end{pmatrix}$
ψ_k	x^{k-1}

Appendix: Algebraic Integers and Characters

In this appendix, we discuss some aspects of character theory that have a different flavor than the material covered in Chapter 6. Logically speaking, this appendix follows Section 15, although here we require a somewhat higher level of algebraic background than in the main part of the book; for instance, some familiarity with Galois theory would be an asset. Throughout, we let G be a finite group.

We say that $x \in \mathbb{C}$ is an *algebraic integer* if there is some monic polynomial $f \in \mathbb{Z}[X]$ such that $f(x) = 0$.

LEMMA 1. The set of rational numbers that are also algebraic integers is exactly the set of ordinary integers.

PROOF. Suppose that $a/b \in \mathbb{Q}$ is an algebraic integer, where a and b are coprime integers with $1 \neq b \in \mathbb{N}$. Then there are integers $c_0, \ldots, c_{n-1}$ for some $n \in \mathbb{N}$ such that

$$\left(\frac{a}{b}\right)^n + c_{n-1}\left(\frac{a}{b}\right)^{n-1} + \ldots + c_1\left(\frac{a}{b}\right) + c_0 = 0.$$

Upon multiplying through by b^n, we see that $a^n \equiv 0 \pmod{b}$; this contradicts the fact that a and b are coprime. ∎

We will make use of the following standard result, whose proof we sketch in the exercises.

PROPOSITION 2. The algebraic integers form a subring of $\mathbb{C}$. ∎

The relevance of algebraic integers to the representation theory of finite groups is established by the following basic fact:

PROPOSITION 3. Let χ be a character of G. Then $\chi(g)$ is an algebraic integer for any $g \in G$.

PROOF. Let $g \in G$. Then $\chi(g)$ is a sum of roots of unity by part (iii) of Proposition 14.4. Any root of unity is an algebraic integer; for instance, if ω is an nth root of unity, then $f(\omega) = 0$, where $f(X) = X^n - 1 \in \mathbb{Z}[X]$. Since the set of algebraic integers is a ring by Proposition 2, it follows that any sum of roots of unity, and in particular $\chi(g)$, is an algebraic integer. ∎

The next result is necessary for both of our intended applications.

LEMMA 4. If χ is an irreducible character of G and $g \in G$, then $|G : C_G(g)|\chi(g)/\chi(1)$ is an algebraic integer.

PROOF. Let S be the simple $\mathbb{C}G$-module having character χ. Let $g \in G$, let K be the conjugacy class of g in G, and let $\alpha \in \mathbb{C}G$ be the class sum $\sum_{x \in K} x$. Consider the map $\varphi : S \to S$ defined by $\varphi(s) = \alpha s$ for $s \in S$. We observed in the proof of Theorem 14.3 that α lies in the center of $\mathbb{C}G$, and from this it follows that $\varphi \in \text{End}_{\mathbb{C}G}(S)$; therefore by Schur's lemma, there is some $\lambda \in \mathbb{C}$ such that $\alpha s = \lambda s$ for all $s \in S$. By taking traces, we now obtain the equation

$$\lambda\chi(1) = \sum_{x \in K} \chi(x) = |K|\chi(g) = |G : C_G(g)|\chi(g).$$

Therefore $\lambda = |G : C_G(g)|\chi(g)/\chi(1)$.

Let $\tau : \mathbb{C}G \to \mathbb{C}G$ be defined by $\tau(z) = z\alpha$ for $z \in \mathbb{C}G$. It follows from the proof of Lemma 13.11 that $\tau \in \text{End}_{\mathbb{C}G}(\mathbb{C}G)$. Now since S is a simple $\mathbb{C}G$-module, we can view S as being a submodule of $\mathbb{C}G$, and for $0 \neq s \in S \subseteq \mathbb{C}G$ we have $\tau(s) = s\alpha = \alpha s = \lambda s$ since α is a central element. Therefore λ is an eigenvalue of τ, and so if we let A be the matrix of τ with respect to the $\mathbb{C}$-basis G for $\mathbb{C}G$, then we now have $\det(\lambda I - A) = 0$. But we see easily that each entry of A is either 0 or 1, and from this it follows that $f(X) = \det(XI - A)$ is a monic polynomial in X with coefficients in $\mathbb{Z}$. Since $f(\lambda) = 0$, we conclude that λ is an algebraic integer. ∎

We can now prove the first main result of this appendix, a fact that was mentioned in Section 14:

PROPOSITION 5. Let χ be an irreducible character of G. Then $\chi(1)$ divides $|G|$.

PROOF. Let $g_1, \ldots, g_r$ be a set of conjugacy class representatives of G. We know for each i that $|G : C_G(g_i)|\chi(g_i)/\chi(1)$ and $\overline{\chi(g_i)}$ are algebraic integers, the former by Lemma 4 and the latter by part (iv) of Proposition 14.4 and Proposition 3. Using row orthogonality, we see that

$$\frac{|G|}{\chi(1)} = \frac{1}{\chi(1)} \sum_{i=1}^{k} |G : C_G(g_i)|\chi(g_i)\overline{\chi(g_i)}$$

$$= \sum_{i=1}^{k} \left(|G : C_G(g_i)| \frac{\chi(g_i)}{\chi(1)} \cdot \overline{\chi(g_i)} \right),$$

which by Proposition 2 is a rational algebraic integer; the result now follows from Lemma 1. ∎

The goal of the remainder of this appendix is to prove Burnside's theorem on the solvability of groups of order $p^a q^b$, which was mentioned in Section 11.

LEMMA 6. Let χ be a character of G, let $g \in G$, and define $\gamma = \chi(g)/\chi(1)$. If γ is a non-zero algebraic integer, then $|\gamma| = 1$.

PROOF. We see from part (iii) of Proposition 14.4 that $\chi(1)\gamma$ is a sum of $\chi(1)$ nth roots of unity, where n is the order of g; therefore $|\gamma| \leq 1$. Suppose that $0 < |\gamma| < 1$, and assume that γ is an algebraic integer. Since γ is an average of d complex roots of unity, the same will be true of any algebraic conjugate of γ. (Here we are applying Galois theory. An algebraic conjugate of γ is the image of γ under any automorphism of K that fixes $\mathbb{Q}$, where K is a suitable extension of $\mathbb{Q}$ containing γ.) In particular, every conjugate of γ has absolute value at most 1, and thus the product of the conjugates of γ has absolute value less than 1. But it follows from [25, Theorem 2.5] that this product is, up to sign, a power of the constant term of the minimal polynomial of γ over $\mathbb{Q}$, a polynomial that by [25, Lemma 2.12] is monic and has integer coefficients. Therefore, the constant term of the minimal polynomial of γ must be zero, which is a contradiction. Hence γ cannot be an algebraic integer. ∎

We now prove a classical theorem due to Burnside.

THEOREM 7. If G has a conjugacy class of non-trivial prime power order, then G is not simple.

PROOF. Suppose that G is simple and that the conjugacy class of $1 \neq g \in G$ is of order p^n, where p is prime and $n \in \mathbb{N}$. (Observe that G must be non-abelian in this event.) From column orthogonality, we obtain

$$0 = \frac{0}{p} = \frac{1}{p} \cdot \sum_{i=1}^{r} \chi_i(g)\chi_i(1) = (1/p) + \sum_{i=2}^{r} \chi_i(g)\chi_i(1)/p,$$

where $\chi_1, \ldots, \chi_r$ are the irreducible characters of G. Since $-1/p$ is by Lemma 1 not an algebraic integer, it follows from Proposition 2 that $\chi_i(g)\chi_i(1)/p$ is not an algebraic integer for some $2 \leq i \leq r$. As $\chi_i(g)$ is an algebraic integer, this implies that $p \nmid \chi_i(1)$ and that $\chi_i(g) \neq 0$. Now $|G : C_G(g)| = p^n$ is coprime to $\chi_i(1)$, so we have $a|G : C_G(g)| + b\chi_i(1) = 1$ for some $a, b \in \mathbb{Z}$. Thus

$$\chi_i(g)/\chi_i(1) = a|G : C_G(g)|\chi_i(g)/\chi_i(1) + b\chi_i(g),$$

which by Proposition 3 and Lemma 4 is an algebraic integer, and therefore $|\chi_i(g)| = \chi_i(1)$ by Lemma 6. Consequently, we see that $g \in Z_i = \{x \in G \mid |\chi_i(x)| = \chi_i(1)\}$. But $Z_i = 1$ by Corollary 15.11, which gives a contradiction. ∎

Finally, we have Burnside's famed $p^a q^b$ theorem:

BURNSIDE'S THEOREM. If $|G| = p^a q^b$, where p and q are primes, then G is solvable.

PROOF. We use induction on $a + b$. If $a + b = 1$, then G has prime order and hence is solvable by Proposition 11.1. Hence we assume that $a + b \geq 2$ and that any group of order $p^r q^s$ is solvable whenever $r + s < a + b$. Let Q be a Sylow q-subgroup of G. If $Q = 1$, then G is a p-group and hence is solvable by Corollary 11.5, so we assume that $Q \neq 1$, in which case $Z(Q) \neq 1$ by Theorem 8.1. Let $1 \neq g \in Z(Q)$. Then as $Q \leqslant C_G(g)$, we have $|G : C_G(g)| = p^n$ for some $n \leq a$. If $n = 0$, then $g \in Z(G)$ and hence G is not simple; and if $n > 0$, then by Proposition 3.13 and Theorem 7 we see that G is not simple. Therefore, G has a non-trivial proper normal subgroup N.

By induction, both N and G/N are solvable, and hence G is solvable by part (iii) of Proposition 11.3. ∎

We observed on page 100 that there are non-solvable groups of order $p^a q^b r^c$, where p, q, and r are distinct primes. In fact, up to isomorphism there are eight such groups. Six of these groups are projective special linear groups. (Recall that A_5 and A_6 are, by Exercises 6.2 and 6.5, isomorphic with projective special linear groups.) The remaining two groups are members of a family of groups called the projective special unitary groups (see [24, p. 245]).

EXERCISES

1. For $x \in \mathbb{C}$, let $\mathbb{Z}[x] = \{f(x) \mid f(X) \in \mathbb{Z}[X]\}$; this is an abelian subgroup of $\mathbb{C}$. Show that x is an algebraic integer iff $\mathbb{Z}[x]$ is finitely generated.

2. (cont.) Show that the set of algebraic integers is a subring of $\mathbb{C}$. (HINT: Use Exercise 1 and the well-known fact (see [1, p. 145]) that any subgroup of a finitely generated abelian group is finitely generated.)

3. Let G be a finite group, and define a function $\psi \colon G \to \mathbb{C}$ by setting $\psi(g) = |\{(x, y) \in G \times G \mid [x, y] = g\}|$ for $g \in G$. Show that

$$\psi = \sum_{i=1}^{r} \frac{|G|}{\chi_i(1)} \chi_i,$$

which in light of Proposition 5 shows that ψ is a character. (Compare with Exercise 3.6.)

Bibliography

[1] Adkins, William A., and Weintraub, Steven H., *Algebra: An Approach via Module Theory*, Springer-Verlag, 1992.

[2] Alperin, J. L., "A classification of n-abelian groups," Can. J. Math. **21** (1969), 1238–1244.

[3] Alperin, J. L., book review, Bull. A.M.S. (N.S.) **10** (1984), 121–123.

[4] Alperin, J. L., *Local Representation Theory*, Cambridge Univ., 1986.

[5] Alperin, J. L., "Brauer's induction theorem and the symmetric groups," Comm. Alg. **15** (1987), 47–51.

[6] Alperin, J. L., "Cohomology is representation theory," in *The Arcata Conference on Representations of Finite Groups*, Amer. Math. Soc., 1987.

[7] Aschbacher, Michael, *Finite Group Theory*, Cambridge Univ., 1986.

[8] Bender, Helmut, and Glauberman, George, *Local Analysis for the Odd Order Theorem*, Cambridge Univ., 1994.

[9] Burton, David H., *Elementary Number Theory*, Allyn and Bacon, 1980.

[10] Collins, M. J., *Representations and Characters of Finite Groups*, Cambridge Univ., 1990.

[11] Feit, Walter, "Theory of finite groups in the twentieth century," in *American Mathematical Heritage: Algebra and Applied Mathematics*, Texas Tech Univ., 1981.

[12] Feit, Walter, and Thompson, John G., "Solvability of groups of odd order," Pacific J. Math **13** (1963), 775–1029.

[13] Fulton, William, and Harris, Joe, *Representation Theory: A First Course,* Springer-Verlag, 1991.

[14] Hall, Philip, "A characteristic property of soluble groups," J. London Math Soc. **12** (1937), 198–200.

[15] Isaacs, I. Martin, *Character Theory of Finite Groups,* Academic Press, 1976 (reprinted by Dover, 1994).

[16] Isaacs, I. Martin, *Algebra: A Graduate Course,* Brooks/Cole, 1994.

[17] Jacobson, Nathan, *Basic Algebra I,* W. H. Freeman, 1985.

[18] Jacobson, Nathan, *Basic Algebra II,* W. H. Freeman, 1989.

[19] Lang, Serge, *Algebra,* Addison-Wesley, 1993.

[20] Mac Lane, Saunders, *Homology,* Springer-Verlag, 1963 (reprinted 1995).

[21] Neumann, Peter M., "A lemma that is not Burnside's," Math. Scientist **4** (1979), 133–141.

[22] Robinson, Derek J. S., *A Course in the Theory of Groups,* Springer-Verlag, 1982 (reprinted 1994).

[23] Rose, John S., *A Course on Group Theory,* Cambridge Univ., 1978 (reprinted by Dover, 1994).

[24] Rotman, Joseph J., *An Introduction to the Theory of Groups,* Springer-Verlag, 1995.

[25] Stewart, I. N., and Tall, D. O., *Algebraic Number Theory,* Chapman and Hall, 1987.

[26] Suzuki, Michio, *Group Theory I,* Springer-Verlag, 1982.

[27] Thompson, John G., "Nonsolvable finite groups all of whose local subgroups are solvable," Bull. A.M.S. **74** (1968), 383–437

List of Notation

General

$\mathbb{N}, \mathbb{Z}$	natural numbers, integers
$\mathbb{Q}, \mathbb{C}$	rational numbers, complex numbers
$B \subseteq A$	B is a subset of A
$B \subset A$	B is a proper subset of A
$A - B$	set of elements of A that are not in B
$x \equiv y \pmod{n}$	$x - y$ is divisible by n

Chapter 1

$H \leqslant G$	H is a subgroup of G
$H < G$	H is a proper subgroup of G
$H \trianglelefteq G$	H is a normal subgroup of G
$H \triangleleft G$	H is a proper normal subgroup of G
1	identity element of G; trivial subgroup of G
$<X>$	subgroup generated by X
$<x_1, \ldots, x_n>$	subgroup generated by $\{x_1, \ldots, x_n\}$
$[x, y]$	$xyx^{-1}y^{-1}$
$|G|$	order of G
$|G : H|$	index of H in G

xH	left coset of H containing x
G/H	coset space of H in G; quotient group of G by H
Σ_X	group of permutations of X
Σ_n	symmetric group of degree n
A_n	alternating group of degree n
$\mathbf{Z_n}$	cyclic group of order n
$\mathbf{Z}$	infinite cyclic group
$\varphi(G)$	image of G under φ
$\mathrm{Aut}(G)$	automorphism group of G
$\mathrm{Inn}(G)$	inner automorphism group of G
$\mathrm{Out}(G)$	$\mathrm{Aut}(G)/\mathrm{Inn}(G)$
$Z(G)$	center of G
$R^\times$	group of units of R
G'	derived group of G
$G_1 \times \ldots \times G_n$	direct product of $G_1, \ldots, G_n$
$N \rtimes H, N \rtimes_\varphi H$	semidirect product of N by H
D_{2n}	dihedral group of order $2n$
Gx	orbit of x under action of G
G_x	stabilizer of x under action of G
$C_G(x)$	centralizer of x in G
$N_G(H)$	normalizer of H in G

Chapter 2

$\mathcal{M}_n(F)$	F-algebra of $n \times n$ matrices over F
$\mathrm{GL}(n, F)$	group of invertible $n \times n$ matrices over F
$\mathrm{GL}(n, q)$	$\mathrm{GL}(n, F)$ when $|F| = q$
$\dim_F(V)$	dimension over F of V
B	Borel subgroup of $\mathrm{GL}(n, F)$
W	Weyl subgroup of $\mathrm{GL}(n, F)$
$X_{ij}(\alpha)$	transvection
$V_n(F)$	space of length n column vectors over F
U	upper unitriangular subgroup of $\mathrm{GL}(n, F)$
T	diagonal subgroup of $\mathrm{GL}(n, F)$
U_P	unipotent radical of a parabolic subgroup P
L_P	Levi complement of U_P
$\mathrm{SL}(n, F)$	group of determinant 1 $n \times n$ matrices over F
$\mathrm{PGL}(n, F)$	$\mathrm{GL}(n, F)/Z(\mathrm{GL}(n, F))$
$\mathrm{PSL}(n, F)$	$\mathrm{SL}(n, F)/Z(\mathrm{SL}(n, F))$

Chapter 4

$G^{(k)}$	kth term of derived series of G
$\mathrm{Aff}(V)$	group of affine transformations of V
$[H, K]$	$<\{[h, k] \mid h \in H,\ k \in K\}>$

Chapter 5

R^{op}	opposite ring of R
$M_1 \oplus \ldots \oplus M_n$	direct sum of modules $M_1, \ldots, M_n$
nM	direct sum of n copies of M
$\mathrm{Hom}_R(M, N)$	set of R-module homomorphisms from M to N
$M \otimes_S N$	tensor product of M with N over S
M^*	$\mathrm{Hom}_R(M, R)$
RG	group ring of G over R
$\mathrm{End}_R(M)$	$\mathrm{Hom}_R(M, M)$
$C^n(H, A)$	normalized n-cochains of (H, A)
$Z^n(H, A)$	n-cocycles of (H, A)
$B^n(H, A)$	n-coboundaries of (H, A)
$H^n(H, A)$	$Z^n(H, A)/B^n(H, A)$
$\mathrm{Ann}(M)$	annihilator of M
$\mathrm{rad}(A)$	radical of A

Chapter 6

$f_1, \ldots, f_r$	dimensions of the simple $\mathbb{C}G$-modules
$\chi_1, \ldots, \chi_r$	irreducible characters of G
χ_U	character of the $\mathbb{C}G$-module U
(α, β)	inner product of class functions α and β
K_χ	$\{x \in G \mid \chi(x) = \chi(1)\}$
Z_χ	$\{x \in G \mid |\chi(x)| = \chi(1)\}$
$\mathrm{Ind}_H^G V$	induction of module V from H to G
$\mathrm{Res}_H^G U$	restriction of module U from G to H
φ^G	induction of character φ from H to G
$\chi|_H$	restriction of character χ from G to H

Index

Graduate Texts in Mathematics

continued from page ii